REACTION
KINETICS
AND
REACTOR DESIGN

PRENTICE-HALL INTERNATIONAL SERIES
IN THE PHYSICAL AND CHEMICAL ENGINEERING SCIENCES

NEAL R. AMUNDSON, EDITOR, *University of Houston*

ADVISORY EDITORS

ANDREAS ACRIVOS, *Stanford University*
JOHN DAHLER, *University of Minnesota*
THOMAS J. HANRATTY, *University of Illinois*
JOHN M. PRAUSNITZ, *University of California*
L. E. SCRIVEN, *University of Minnesota*

AMUNDSON *Mathematical Methods in Chemical Engineering: Matrices and Their Application*
ARIS *Introduction to the Analysis of Chemical Reactors*
BALZHISER, SAMUELS, AND ELIASSEN *Chemical Engineering Thermodynamics*
BRIAN *Staged Cascades in Chemical Processing*
BUTT *Reaction Kinetics and Reactor Design*
DENN *Process Fluid Mechanics*
DENN *Stability of Reaction and Transport Processes*
DOUGLAS *Process Dynamics and Control: Vol. 1, Analysis of Dynamic Systems*
DOUGLAS *Process Dynamics and Control: Vol. 2, Control System Synthesis*
FOGLER *The Elements of Chemical Kinetics and Reactor Calculations: A Self-Paced Approach*
FREDRICKSON *Principles and Applications of Rheology*
FRIEDLY *Dynamic Behavior of Processes*
HIMMELBLAU *Basic Principles and Calculations in Chemical Engineering*, 3rd edition
HOLLAND *Fundamentals and Modeling of Separation Processes: Absorption, Distillation, Evaporation, and Extraction*
HOLLAND *Multicomponent Distillation*
HOLLAND *Unsteady State Processes with Applications in Multicomponent Distillation*
HOLLAND AND ANTHONY *Fundamentals of Chemical Reaction Engineering*
LEVICH *Physicochemical Hydrodynamics*
MEISSNER *Processes and Systems in Industrial Chemistry*
MODELL AND REID *Thermodynamics and its Applications*
MYERS AND SEIDER *Introduction to Chemical Engineering and Computer Calculations*
NEWMAN *Electrochemical Systems*
PERLMUTTER *Stability of Chemical Reactors*
PRAUSNITZ *Molecular Thermodynamics of Fluid-Phase Equilibria*

REACTION
KINETICS
AND
REACTOR DESIGN

JOHN B. BUTT

Department of Chemical Engineering
Northwestern University
Evanston, Illinois

PRENTICE-HALL, INC.,
Englewood Cliffs, New Jersey 07632

Library of Congress Cataloging in Publication Data

BUTT, JOHN B   date
    Reaction kinetics and reactor design.

    Includes index.
    1. Chemical reaction, Rate of.   2. Chemical
reactors.   I. Title.
QD502.B87 1980   541'.39   79–16115
ISBN   0–13–753335–7

Editorial/production supervision and interior design
by *Virginia Huebner*
Cover design by Edsal Enterprises
Manufacturing buyer: *Gordon Osbourne*

PRENTICE-HALL INTERNATIONAL, INC., *London*
PRENTICE-HALL OF AUSTRALIA PTY. LIMITED, *Sydney*
PRENTICE-HALL OF CANADA, LTD., *Toronto*
PRENTICE-HALL OF INDIA PRIVATE LIMITED, *New Delhi*
PRENTICE-HALL OF JAPAN, INC., *Tokyo*
PRENTICE-HALL OF SOUTHEAST ASIA PTE. LTD., *Singapore*
WHITEHALL BOOKS LIMITED, *Wellington, New Zealand*

*To all my students
who, over the years,
have taught me at least
as much as I have taught them.*

Contents

Preface

It is probably obvious even to the beginning student that much of chemical engineering is centered about problems involving chemical transformation, that is, chemical reaction. It is probably not so obvious, at least in the beginning, that the rate at which such transformations occur is the determining factor in a great number of the processes that have been developed over the years to produce that vast array of goods that we consider an integral part of contemporary life. The study, analysis, and interpretation of the rates of chemical reactions is, itself, a legitimate field of endeavor. It ranges in scope from those problems concerned with the fundamentals of detailed mechanisms of chemical transformation and the associated rates to problems that arise during the development and implementation of procedures for chemical reactor and process design on a large scale. If we must give names to these two extremes, we might call the first "chemical kinetics", and the second "chemical reaction engineering".

In the following we shall range from one limit to the other, although our primary objective is the understanding of kinetic principles and their application to engineering problems. What will we find? For one thing, we will find that chemical reactions are not simple things; those fine, balanced equations which everyone has used in solving stoichiometry problems ordinarily represent only the sum of many individual steps. We will find that the rates of chemical transformation, particularly in engineering application, are often affected by the rates of other processes, such as the transport of mass or heat, and cannot be isolated from the physical enviroment. We will find that the normal dependence of reaction rate on temperature is one of the most intractable of nonlinearities in nature, providing at the same time many of the difficulties and many of the challenges in the analysis of chemical rates. We will find that often it is not the absolute rate of a single reaction but the relative rates of two or more reactions that will be important in determining a design. We will find that space as well as time plays an important role in reaction engineering, and in the treatment of such problems it will be necessary to develop some facility in the use of rational mathematical models. Finally, we will find that the artful compromise is as important, if not more so, in our applications of reaction kinetics as it is in all the other areas of chemical engineering practice.

This may seem like a very short list of what is to be found if the topic is truly as important as we have indicated. It is intentionally short, because the essence should not be submerged in detail quite so soon, or, to paraphrase Thomas a Kempis, it is better not to speak a word at all than to speak more words than we should.

The material of this text is intended primarily to provide instruction at the undergraduate level in both chemical kinetics and reactor design. Of particular concern has been the detailing of reaction kinetics beyond phenomonological description. The rationale for the Arrhenius equation was a personal mystery to the author in earlier years, who hopes an appropriate solution is revealed in Chapter 2. Numerous other aspects of classical theories of chemical kinetics are assembled Chapter 2 and Chapter 3 to give some perception of the origin of phenomological rate laws and an understanding of the differing types of elementary reaction steps. In Chapter 4 we swap the beret of the theoretician for the hard hat of the engineer, in pursuit of means for developing rational chemical reactor design and analysis models. A parallelism between mixing models and reactor models has been maintained in order to demonstrate clearly how reaction kinetic laws fit into reactor design. Chapters 4 through 6 are based on homogeneous models, and proceed from standard plug flow and stirred tank analysis to description of nonideal behavior via dispersion, segregated flow, mixing cell and combined model approaches. Phenomena associated with reaction in more than one phase are treated in Chapter 7 but no attempt is made to develop multiphase reactor models. The fact that reaction selectivity as well as reaction rate is an important and often determining factor in chemical reaction or reactor analysis is kept before the eyes of the reader throughout the text.

The exercises are an intentionally well-mixed bag. They range from simple application of equations and concepts developed in the text to relatively open-ended situations which may require arbitarary judgement and, in some instances, have no unique answer. The units employed are equally well mixed. Historically, multiple systems of measure have been a curse of the engineering profession and such is the case here particularly, where we range from the scientific purity of Planck's constant to the ultimate practicality of a barrel of oil. The SI system will eventually provide standarization, it is to be hoped, but this is not a short term proposition. Because both author and reader must continue to cope with diverse sets of units, no attempt at standarization has been made here.

Symbols are listed in alphabetical order by the section of the chapter in which they appear. Only symbols which have not been previously listed or which are used in a different sense from previous listings are included for each section. Symbols used in equations for simplification of the form are generally defined immediately thereafter and are not listed in Notation.

Each chapter is divided into more or less self-contained modules dealing with a unified concept or a group of related concepts. Similarly, the exercises and notation are keyed to the individual modules, so that a variety of possibilities exist for pursuit of the material presented.

Acknowledgment must be made to teachers and colleagues who, over the years, have had influence in what is to be found in this text. I am grateful to the late Charles E. Littlejohn and R. Harding Bliss, to Professors C. A. Walker, H. M. Hulburt, and R. L. Burwell, Jr., and especially to Professor C. O. Bennett, who offered many constructive and undoubtedly kind comments during preparation of this manuscript. Tanks also to R. Mendelsohn, D. Casleberry and J. Pherson for typing various sections of the manuscript, and to the Northwestern chemical engineering students for detecting unworkable problems, inconsistent equations, and all the other gremlins waiting to smite the unwary author.

JOHN B. BUTT

Evanston, Illinois

Apparent Reaction Kinetics in Homogeneous Systems

1

1.1 Mass Conservation and Chemical Reaction

Certainly, the most fundamental of laws governing the chemical transformations and separations with which chemical engineering is involved is that of the *conservation of mass*. Consider the steady-state separation process depicted on Figure 1.1. A stream, L, mass/time, containing two components, A and

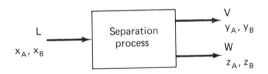

L
x_A, x_B → Separation process → V y_A, y_B / W z_A, z_B

Figure 1.1 Simple separation

B, is fed to a separation process of some sort which divides in into two product streams, V and W, mass/time, also containing components A and B. We can define the mass fractions of components A and B in L, V, and W as x_A, x_B, y_A, y_B, and z_A, z_B, respectively. Mass is conserved in this separation; we may express this mathematically with the following simple relations:

$$L = V + W \tag{1-1}$$

$$Lx_A = Vy_A + Wz_A \tag{1-2}$$

$$Lx_B = Vy_B + Wz_B \tag{1-3}$$

1

Now, since we know also that each stream consists of the sum of its parts, then

$$x_A + x_B = 1 \qquad (1\text{-}4)$$

$$y_A + y_B = 1 \qquad (1\text{-}5)$$

$$z_A + z_B = 1 \qquad (1\text{-}6)$$

This, in turn, means that only two of the three mass-balance relationships (1-1) to (1-3) are independent and can be used to express the law of conservation of mass for the separation. We are then left with a system of five equations and nine potential unknowns such that if any four are specified, the remaining five may be determined. Of course, all we have done is to say:

total mass in/time = total mass out/time

mass A in/time = mass A out/time

mass B in/time = mass B out/time

Since the uniform time dimension divides out of each term of these equation, our result is the direct mass-conservation law.

Now let us consider a slightly different situation in which the process involved is not a separation but a chemical transformation. In fact, we shall simplify the situation to a single input and output stream as in Figure 1.2,

Figure 1.2 Chemical reaction process

with a feed stream consisting of component A alone. However, within the process a chemical reaction occurs in which B is formed by the reaction A → B. If the reaction is completed within the process, all the A reacts to form B and mass conservation requires that the mass of B produced equal the mass of A reacted. The material balance is trivial:

$$L = W \qquad (x_A = 1, z_B = 1) \qquad (1\text{-}7)$$

What happens, though, if not all of the A reacts to form B in the process? Then, obviously, the mass of B leaving is not equal to the mass of A entering, but rather

$$L = W z_A + W z_B \qquad (1\text{-}8)$$

This chemical reaction process illustrates in a simplistic way the general concerns of this text, which are to determine z_A and z_B given a certain type and size of reaction process, or to determine the type and size of reaction process needed to produce a specified z_A and z_B.

Two factors enter into this problem. The first is the *stoichiometry* of the reaction transforming A to B. Chemical equations as normally written express the relationship between molal quantities of reactants and products, and it is necessary to transform these to the mass relationship in problems of mass conservation involving chemical reaction. The details of this are clear from the familiar combustion/mass balance problems, which seem so vexing the first time they are encountered. In our simple example reaction, the stoichiometric relationship is $1:1$, so mass conservation requires the molecular weight of B to equal that of A, and thus the mass of B produced equal the mass of A reacted. This particular type of reaction is called an *isomerization reaction* and is a common and important reaction in industrial application.

The second factor is the *rate* at which A reacts to form B. Consider the problem in which we wish to determine z_A and z_B, given the type and size of process used in Figure 1.2. For the mass balance on component A, we want to write

$$\text{mass A in/time} = \text{mass A out/time} + \text{mass A reacted/time} \qquad (1\text{-}9)$$

and for B (recalling that there is no B in the feed)

$$\left(\frac{\text{mass B out}}{\text{time}}\right) = \left(\frac{\text{mass B formed}}{\text{time}}\right) = \left(\frac{\text{mass A reacted}}{\text{time}}\right) \qquad (1\text{-}10)$$

The last two terms of equation (1-10) incorporate the information concerning the stoichiometric relationship involved in the chemical reaction, since we have already seen that the mass relationship as well as the molal relationship in this particular example is $1:1$. We can also paraphrase the statement of equation (1-10) to say that the *rate* at which B passes out of the system is equal to the *rate* at which it is formed, which is also equal to the *rate* at which A reacts. Thus, the rate of reaction is closely involved in this mass-balance relationship—exactly how is what we are to learn—so for the moment our simple example must remain unsolved.

The most important matter is to define properly a rate of reaction. One undesirable property exhibited by the rate involved in equation (1-10) is that its magnitude depends on the magnitude of the process; and we could have an infinity of processes, as in Figure 1.2, of differing sizes under identical conditions of operation, each producing the same z_A and z_B, yet each having a different reaction rate. *Volume* is the measure of the size of a system, so if we define a reaction rate with respect to unit volume of reaction mixture, the size dependency will be removed. Further, we must be sure to define the rate of reaction such that it will properly reflect the influence of state variables such as composition and temperature but will not be dependent on the particular process or reactor in which the reaction takes place. In accord with this, we can write

$$r' = \text{rate of reaction} = \frac{\text{change in mass of reactant or product}}{\text{time–volume reaction mixture}} = \frac{1}{V}\frac{dM}{dt}$$

$$(1\text{-}11a)$$

or, in molal quantities,

$$r = \text{rate of reaction} = \frac{\text{change in moles of reactant or product}}{\text{time-volume reaction mixture}} = \frac{1}{V}\frac{dN}{dt}$$

$$(1\text{-}11\text{b})$$

where V is the volume of the reaction mixture and M and N are mass and moles, respectively. The rate is positive for change in product and negative for change in reactant.

There are a number of traps involved even in this simple definition. The use of the reference volume as that of the reaction mixture is necessary to account for the fact that in some cases the total volume will change in proportion to the molal balance between reactants and products. If the reaction were, for example, A → 2B and involved ideal gases with no change in temperature or total pressure, the volume of product at the completion of reaction would be twice that of initial reactant. To conserve our definition of reaction rate independent of the size of the system, the volume change must be accounted for. This is not difficult to do, as we shall show a little later. The problem is that one often sees reaction rates written in the following form:

$$r = \frac{dC}{dt} = \frac{d(N/V)}{dt}$$

$$(1\text{-}12)$$

where concentration C (in moles/volume) is used in the rate definition. As the last term in equation (1-12) shows, such a simplification is possible only when the volume of the reaction mixture does not change as the reaction progresses. In most cases it is convenient to use concentration units, however, so one can avoid problems with changes in reference volume if the rate is written

$$r = \frac{1}{V}\frac{d(CV)}{dt}$$

$$(1\text{-}13)$$

where $N = CV$,

A second and not so obvious trap involved in the rate definition is the use of the time derivative, (dM/dt) or (dN/dt). This implies that things are changing with time and may tempt one to associate the appearance of reaction-rate terms in mass conservation equations with unsteady-state processes. This is not necessarily true; in general, one must make a distinction between the ongoing time of operation of some process (i.e., that measured by an observer) and individual phenomena such as chemical reactions which occur at a *steady* rate in an operation that does not vary with time (i.e., steady state). Later we will encounter both steady- and unsteady-state types of processes involving chemical reaction, to be sure, but this depends on the process itself; the reaction-rate definition has been made without regard to a particular process.

Returning for a moment to the mass-conservation equation (1-9), we may now restate it in terms of the definition of the rate of reaction. We use as a basis a specified volume element δ of reaction mixture within the process, for which, under steady conditions,

$$\left(\frac{\text{mass A entering } \delta}{\text{time}}\right) = \left(\frac{\text{mass A leaving } \delta}{\text{time}} + \frac{\text{mass A reacted in } \delta}{\text{time}}\right) \quad (1\text{-}14)$$

as shown in Figure 1.3a. The reaction term in equation (1-14) now conforms to the reaction-rate definition; the left side of the equality is the *input* term and the right side the *output*. It is understood that A refers to a reaction or product species, not an element.

Unit volume element of reaction mixture, δ

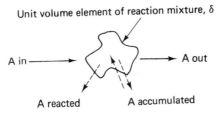

A in ⟶   ⟶ A out

A reacted   A accumulated

Figure 1.3a Mass balance with chemical reaction on a volume element of reaction mixture

In the event of unsteady-state operation, the means of incorporating the rate expression in the mass balance are still the same, but the balance becomes

$$\left(\frac{\text{mass A entering } \delta}{\text{time}}\right) - \left(\frac{\text{mass A leaving } \delta}{\text{time}} + \frac{\text{mass A reacted in } \delta}{\text{time}}\right)$$
$$= \left(\frac{\text{change of mass of A in } \delta}{\text{time}}\right) \quad (1\text{-}15)$$

The input/output terms remain the same, but an *accumulation* term on the right side represents the time dependence of the mass balance.

From equation (1-15) we can immediately derive expressions relating the reaction rate to the type of process for two limiting cases. First, if the volume element δ does not change as the reaction proceeds, and there is no flow of reactant or product into or out of δ, equation (1-15) becomes

$$\left(\frac{\text{mass A reacted in } \delta}{\text{time}}\right) = \left(\frac{\text{change of mass of A in } \delta}{\text{time}}\right)$$

The left-hand side is the definition of the reaction rate of A, r'; and the right-hand side (recalling that δ does not change) is given by the rate of change of concentration of A, so

$$r'_A = \frac{dC_A}{dt} \quad (1\text{-}12)$$

This is the equation for a *batch reactor*. An identical expression applies for molal units.

In the second case we also assume that the volume element δ does not change and that there is no flow of A into or out of δ. However, now the

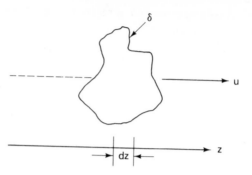

Figure 1.3b Traveling batch reactor

volume element is moving, with others, in some environment such that the distance traversed from some reference point is a measure of the time the reaction has been occurring, as shown in Figure 1.3b. Now, for this little traveling batch reactor we may rewrite the time derivative in terms of length and linear velocity. For a constant velocity, u, as the volume element moves through the length, dz,

$$dt = \frac{dz}{u}$$

and equation (1-13) becomes

$$u\frac{dC_A}{dz} = r'_A \tag{1-16}$$

This is the equation for a *plug-flow reactor*. One must be careful in the use of molal or mass units here. For liquids mass units are usually appropriate and equation (1-16) is correct as written; however, for an ideal gas, u is the molal average velocity and molal units must be employed. These points will be discussed in more detail subsequently.

The batch reactor and the plug-flow reactor are two important types of idealized reactor models which are extremely important in the analysis of chemical reaction processes. Indeed, the origins and applications of such idealized reactor models form the substance of chapter 4.

1.2 Reaction-Rate Equations— The Mass Action Law

While a proper definition of the rate of reaction is necessary, we cannot do much with it until we find how the rate depends on the state variables of the reaction system such as temperature, total pressure, and composition. In chapter 2 we will see that there are rather substantial theoretical reasons for using the forms of rate equation that we do, but for now we shall present these forms in a priori fashion, with the primary objective of familiarization with their properties and manipulation.

The fundamental expression for the rate of an irreversible reaction

$$aA + bB + cC + \ldots \longrightarrow lL + mM + \ldots \tag{I}$$

can generally be written

$$r = kC_A^p C_B^q C_C^r \tag{1-17}$$

which form follows directly from the *law of mass action*. The rate constant k normally depends on the temperature, and the exponents p, q, and r may also be temperature-dependent. These exponents are termed the *order* of the reaction; total order of the reaction is $(p + q + r)$, and order with respect to an individual reactant, say A, is p. The rate constant is a positive quantity, so if the rate of equation (1-17) refers to disappearance of a reactant, it is necessary to include the minus sign. If we rewrite equation (1-17) to refer specifically to the disappearance of A, and use the rate definition of equation (1-12):

$$-kC_A^p C_B^q C_C^r \ldots = -k\left(\frac{N_A}{V}\right)^p \left(\frac{N_B}{V}\right)^q \left(\frac{N_C}{V}\right)^r \ldots \tag{1-18}$$

$$= r_A = \frac{\text{moles A reacted}}{\text{time-volume}}$$

If the total volume of the reaction system does not change, then

$$\frac{dC_A}{dt} = -kC_A^p C_B^q C_C^r \ldots = r_A = \frac{\text{moles A reacted}}{\text{time-volume}} \tag{1-19}$$

In many cases we shall see that the individual orders p, q, and r have numerical values which are the same as the stoichiometric coefficients of the reaction, a, b, and c. In this event it is said that *order corresponds to stoichiometry*; there is, in general, however, no restriction on the numerical values of p, q, and r.

If the reaction is reversible, A, B, C, $\ldots$ can be formed from the products L, M, $\ldots$ and a corresponding form of the law of mass action applies. For this case equation (1-18) becomes

$$-k_f C_A^p C_B^q C_C^r + k_r C_L^u C_M^w = r_A \tag{1-20}$$

Here u and w are the orders of the reverse reaction with respect to L and M, k_f the forward rate constant, and k_r the reverse rate constant. The same comments as before pertain to the relationship between the apparent orders and the stoichiometric coefficients l and m.

The rate constants here have been defined in molal terms and we will follow this convention for the time being, although there is no reason that mass units could not be used, following equation (1-11). The rate expression for the appearance or disappearance of a given component, however, implies a corresponding definition of the rate constant. In equation (1-18) k defines the moles of A reacted per time per volume, and whether the same numerical value of k will also give the moles of B reacted per time per volume depends on the relative values of the stoichiometric coefficients a and b; yes only if $a = b$. Rewrite the reaction as

$$A + \left(\frac{b}{a}\right)B + \left(\frac{c}{a}\right)C \ldots \longrightarrow \left(\frac{l}{a}\right)L + \left(\frac{m}{a}\right)M + \ldots$$

Clearly, (b/a) moles of B react for each mole of A, and so on. The rate of reaction of any constituent, i, is related to the rate of reaction of A by

$$r_i = r\left(\frac{\nu_i}{\nu_A}\right)$$

where ν are the individual stoichiometric coefficients, positive for products and negative for reactants, and $r = -r_A$.

For reversible reactions it is sometimes convenient to consider the individual terms in rate expressions such as equation (1-20). The *forward* rate for A is

$$(r_A)_f = -k_f C_A^p C_B^q C_C^r \ldots \tag{1-20a}$$

and the *reverse* rate

$$(r_A)_r = k_r C_L^u C_M^w \ldots \tag{1-20b}$$

and the *net* rate the algebraic sum of the two:

$$r_A = (r_A)_f + (r_A)_r \tag{1-21}$$

When the forward and reverse rates are equal, the reaction is at equilibrium and we have the result

$$k_f C_A^p C_B^q C_C^r = k_r C_L^u C_M^w \tag{1-22}$$

or

$$\frac{k_f}{k_r} = \frac{C_L^u C_M^w}{C_A^p C_B^q C_C^r} \tag{1-23}$$

It is tempting to say that such a result leads us immediately to the equilibrium constant of the reaction, but this is not generally true. If the reaction under consideration is truly an *elementary step* in the sense we shall discuss later, the ratio (k_f/k_r) does represent the chemical equilibrium constant. But many, even most, chemical transformations which we observe on a macroscopic scale and which obey overall stoiciometric relationships as given by (I) consist of a number of individual or elementary steps which for one reason or another are not directly observable. In such cases observation of the stationary condition of equation (1-22) does not directly yield the equilibrium constant of the reaction, since we have in essence used a simplified model to represent the net result of what may be a complex and interrelated sequence of events. Nonetheless, reaction-rate expressions corresponding to the law of mass action can almost always be applied to homogeneous reactions; Table 1.1a gives a representative selection for various reactions.

TABLE 1.1a
MASS-ACTION-LAW RATE EQUATIONS (II)

Reaction	Rate Law
(1) $CH_3CHO \longrightarrow CH_4 + CO$	$k[CH_3CHO]^{1.5}$
(2) $N_2 + 3H_2 \longrightarrow 2NH_3$	$k[N_2][H_2]^{2.25}[NH_3]^{-1.5}$
(3) $4PH_3 \longrightarrow P_4 + 6H_2$	$k[PH_3]$
(4) $CH_3COCH_3 + HCN \rightleftharpoons (CH_3)_2C{<}^{OH}_{CN}$	$k_f[HCN][CH_3COCH_3]$ $- k_r[(CH_3)_2COHON]$
(5) $(C_2H_5)_2O \longrightarrow C_2H_6 + CH_3CHO$	$k[C_2H_5OC_2H_5]$
(6) $CH_3OH + CH_3CHOHCOOH$ $\longrightarrow CH_3CHOHCOOCH_3 + H_2O$	$k[CH_3OH][CH_3CHOHCOOH]$

It will be convenient in this discussion to maintain a distinction between the elementary steps of a reaction and the overall reaction under consideration. The direct application of the law of mass action where the orders and the stoichiometric coefficients correspond will normally pertain only to the elementary steps of a reaction, as will the dependence of rate on temperature, which we discuss in the next section. Also, the theoretical background to be given later concerning pressure, concentration, and temperature dependence of rate pertains to these elementary steps. Finally, for the forward and reverse steps of a reaction which is an elementary step, the principle of *microscopic reversibility* applies. This states that the reaction pathway most probable in the forward direction is also most probable in the reverse direction. In terms of energy, one may say that the forward and reverse reactions of the elementary step are confronted with the same energy barrier.

1.3   Temperature Dependence of Reaction Rate

Mass-action-law rate equations are sometimes referred to as *separable* forms because they can be written as the product of two factors, one dependent on temperature and the other not. This can be illustrated by writing equation (1-18) as

$$-r_A = k(T)C_A^p C_B^q C_C^r \ldots \tag{1-24}$$

where the rate constant $k(T)$ is indicated to be a function of temperature, as we have mentioned before, and the concentration terms are independent of that variable. The possible dependence of p, q, and r on temperature is small and normally arises from factors associated with the fitting of rate forms to kinetic data—not the problem we are concerned with here.

The temperature dependence of k associated with the rate of an elementary step is almost universally given by the awkward exponential form called the *Arrhenius equation*:

$$k(T) = k°e^{-E/RT} \tag{1-25}$$

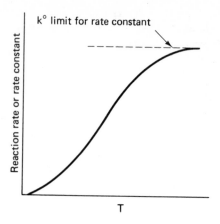

Figure 1.4 Reaction-rate and rate-constant dependence on temperature according to the Arrhenius law

where E is the activation energy of the reaction, T the absolute temperature and $k°$ the preexponential factor. In equation (1-25) $k°$ is written as independent of temperature; in fact, this may not be so, but the dependency of $k°$ on temperature is weak and the exponential term is by far the predominate one.[1] Figure 1.4 illustrates the general form of the dependence of reaction-rate constants obeying the Arrhenius law.

The determination of activation energy from data on the temperature dependence of velocity constant is most often accomplished by graphical analysis of equation (1-25). Taking logarithms of both sides:

$$\ell n\, k(T) = -\frac{E}{RT} + \ell n\, k°$$ (1-25a)

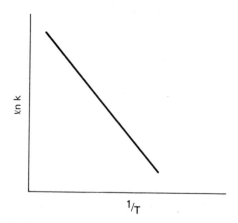

Figure 1.5 Form of Arrhenius plot for determination of activation energy

[1]This point is discussed in exquisite detail in chapter 2.

Thus, a plot of the logarithm of the rate constant versus the reciprocal of temperature is linear with slope $-(E/R)$ and intercept $\ell n\ k°$, as shown in Figure 1.5. In some cases the linear correlation may not be obtained. This may be due to several factors, the most frequent of which are:

1. The mechanism of the reaction changes over the temperature range studied.
2. The form of rate expression employed does not correspond to the reaction occurring (i.e., some composite rate is being correlated).
3. Other rate processes, such as mass diffusion, are sufficiently slow to obscure the reaction rates.
4. The temperature dependence of the preexponential factor, $k°$, becomes important (see problem 10).

In general, the Arrhenius correlation is probably better obeyed than any other in kinetics if properly applied. A problem of particular interest and importance to chemical engineers is that listed as factor 3 above, which we shall discuss in detail in chapter 7.

Stated from a somewhat different point of view, the Arrhenius equation serves to define the activation energy E:

$$\frac{d(\ell n\ k)}{dT} = \frac{E}{RT^2} \qquad (1-26)$$

The form of equation (1-26) is reminiscent of the *van't Hoff relationship* for the temperature dependence of the equilibrium constant[1]:

$$\frac{d(\ell n\ K)}{dT} = \frac{\Delta H°}{RT^2} \qquad (1-27)$$

Consider the reaction $A + B \rightleftarrows L + M$, which is an elementary step. Then the ratio of forward to reverse rate constants does give the equilibrium constant, K, and we may write

$$\frac{d(\ell n\ K)}{dT} = \frac{d[\ell n\ (k_f/k_r)]}{dT} = \frac{\Delta H°}{RT^2} \qquad (1-28)$$

If we arbitrarily divide the total enthalpy change $\Delta H°$ into the difference between two parts, one associated with the forward reaction, $\Delta H_f°$, and the other with the reverse, $\Delta H_r°$, then

$$\frac{d(\ell n\ k_f)}{dT} - \frac{d(\ell n\ k_r)}{dT} = \frac{\Delta H°}{RT^2} = \frac{\Delta H_f°}{RT^2} - \frac{\Delta H_r°}{RT^2} \qquad (1-29)$$

If, further, we arbitrarily associate corresponding terms on the two sides of equation (1-29), then

$$\frac{d(\ell n\ k_f)}{dT} = \frac{\Delta H_f°}{RT^2} \qquad (1-30)$$

[1] In modern terminology, this would be better expressed in terms of $\Delta G°$, the change in Gibbs free energy.

for example, and the enthalpy change ΔH_f corresponds to what was called the activation energy in equations (1-25) and (1-26). This is a rather arbitrary way to associate some physical significance with the activation energy, but the thermodynamic argument above is that originally used to develop equation (1-25). We shall develop a more satisfying theoretical approach both to the form of the Arrhenius equation and the significance of the activation energy in chapter 2. The exponential dependence of rate on temperature makes problems involving reaction kinetics in nonisothermal systems some of the most interesting and occasionally the most difficult to be found in chemical engineering.

1.4   Rate Laws and Integrated Forms for Elementary Steps

a. Individual and overall reactions

We have made a distinction between an overall reaction and its elementary steps in discussing the law of mass action and the Arrhenius equation. Similarly, the basic kinetic laws treated in this section can be thought of as primarily applying to elementary steps. What relationships exist between these elementary steps and the overall reaction? In Table 1.1a we gave as an example of a mass action rate law that for the decomposition of diethyl ether:

$$C_2H_5OC_2H_5 \longrightarrow C_2H_6 + CH_3CHO \qquad (III)$$

for which

$$-r = k[C_2H_5OC_2H_5] \qquad (1\text{-}31)$$

The overall reaction written in (III) does not indicate how the transformation from reactants to products actually occurs; indeed, (III) is quite misleading in this regard and, even worse, the fact is camouflaged by the agreement of order and stoichiometry in equation (1-31). The actual sequence of reaction is

$$
\begin{array}{ll}
C_2H_5OC_2H_5 \longrightarrow {\cdot}CH_3 + {\cdot}CH_2OC_2H_5 & (a) \\
{\cdot}CH_3 + C_2H_5OC_2H_5 \longrightarrow C_2H_6 + {\cdot}CH_2OC_2H_5 & (b) \\
{\cdot}CH_2OC_2H_5 \longrightarrow {\cdot}CH_3 + CH_3CHO & (c) \\
{\cdot}CH_3 + {\cdot}CH_2OC_2H_5 \longrightarrow C_2H_5OC_2H_5 & (d)
\end{array}
\qquad (IV)
$$

where each of the reactions (IVa) to (IVd) represents an elementary step of the overall decomposition reaction. In practice, the intermediate species, such as ${\cdot}CH_3$ and ${\cdot}CH_2OC_2H_5$, are very reactive and consequently have very short lifetimes and appear at very low concentrations, a point that will be discussed in detail later. They are, therefore, not readily observable, in contrast to the long-lived products C_2H_6 and CH_3CHO. The reader may verify that the sum of the elementary steps of (IV) gives the overall reaction (III).

The ether decomposition is an example of a *chain reaction*, in which a cycle of elementary steps such as (IVb) and (IVc), which involve the active intermediates in the chain, produce the final products. There are formal methods for treating the kinetics of such chain reactions which we will encounter later; the important point here is to note the relationship between the overall reaction and the elementary steps and to note that even though the overall kinetics are apparently in accord with those for an elementary step, reaction (III) is not one.

As a second example consider a heterogeneous reaction—one occurring in two phases—rather than the homogeneous cases we have been discussing so far. The water gas reaction is a well-known example of a reaction whose rate is influenced by the presence of a solid catalyst:

$$H_2O + CO \xrightarrow{\text{catalyst}} H_2 + CO_2 \qquad \text{(V)}$$

where the two phases involved are the gaseous reaction mixture and the solid catalytic surface. It is more apparent for (V) than for (III) that the overall reaction must consist of some sequence of elementary steps, since (V) in no way accounts for the influence of the catalyst on the reaction. If we let S represent some chemically active site on the catalytic surface, reaction (V) can be written

$$
\begin{aligned}
H_2O + S &\longrightarrow H_2 + SO \qquad &\text{(a)} \\
CO + SO &\longrightarrow CO_2 + S \qquad &\text{(b)}
\end{aligned}
\qquad \text{(VI)}
$$

where SO is an oxide complex on the surface and plays the role of the intermediate in this reaction sequence.

When we compare (IV) and (VI) it is clear that much more detail concerning elementary steps is given in the former. The two reactions of (VI) provide a closer description of the water gas catalysis than does (V) but are not necessarily themselves the elementary steps. This is important in the applications of kinetics that we are concerned with, since frequently it is necessary to work with only partial information and knowledge of complicated reactions. The two-step sequence of (VI) is more desirable than (V), since it provides a means for incorporating the catalytic surface into the reaction scheme even though the two steps involved may not be elementary. We may summarize this by saying that the two-step sequence provides for an *essential feature* of the reaction, the involvement of the surface, which is absent in the overall reaction (V). Applied kinetics often requires the modeling of complicated reactions in terms of individual steps which incorporate the essential features of the overall reaction. As one develops more detailed information concerning a reaction, it will be possible to write in more detail the sequences of reaction steps involved, approaching (ideally) the actual elementary steps.

Boudart (M. Boudart, *Kinetics of Chemical Processes*, Prentice-Hall, Englewood Cliffs, N.J., 1968) has given a convenient means of classification of reaction sequences such as (IV) and (VI). The intermediates, such as

$\cdot CH_3$, S, and SO, are *active centers*, since the reaction proceeds via steps involving the reactivity of these species. In some cases, such as (IV) and (VI), the active centers are reacted in one step and regenerated in another [$\cdot CH_3$ and $\cdot CH_2OC_2H_5$ in (IV), S and SO in (VI)], so that a large number of product molecules can be produced through the action of a single active center. This is termed a *closed* sequence. In the case where this utilization and regeneration of active centers does not occur, the sequence is termed an *open* one and a single active center is associated with a single-product molecule-producing step. The gas-phase decomposition of ozone is an example of such an open sequence:

$$2O_3 \longrightarrow 3O_2 \tag{VII}$$

where the individual steps are

$$O_3 \longrightarrow O_2 + O \tag{a}$$
$$O + O_3 \longrightarrow O_2 + O_2 \tag{b}$$

(VIII)

and oxygen atoms are the active center for the reaction.

We shall ultimately be concerned with methods for establishing the rate laws for overall reactions such as (IV), (VI), and (VII) from a knowledge of (or speculation as to) the elementary steps. To do this we must certainly develop some facility with the simple rate laws that elementary reactions might be expected to obey. This is the topic of the following section.

b. Rates and conversions for simple reactions

In this section we must be careful to respect our prior concern regarding the definition of rate with regard to the volume of reaction mixture involved. Further, since we wish to concentrate attention on the kinetics, we shall study systems in which the conservation equation contains the reaction term alone, which is the batch reactor of equation (1-12). It is convenient to view this type of reactor in a more general sense, as one in which all elements of the reaction mixture have been in the reactor for the same length of time. That is, all elements have the same *age*. Since the reactions we are considering here occur in a single phase, the relationships presented below pertain particularly to *homogeneous batch* reactions, and temperature changes are absent, so the systems are isothermal.

The most simple of these reactions is the class in which the rates are irreversible and the reaction system is one of constant volume. The most important of these are zero-, first-, and second-order with respect to the reactant(s), respectively.

Zero Order. It may seem somewhat at odds with the law of mass action to talk about rates which are independent of the concentration of reactant, but apparent *zero-order* reactions do occur, particularly in the description of the overall kinetics of some closed reaction sequences. We consider the model reaction,

$$A \longrightarrow B$$

with the zero-order rate law in molal units as

$$r_A = \frac{1}{V}\frac{dN_A}{dt} = -k = \frac{dC_A}{dt}$$

and in the batch reactor $C_A = C_{A_0}$ at the start of the reaction, $t = 0$. The concentration at any time is

$$\int_{C_{A_0}}^{C_A} dC_A = -\int_0^t k\, dt \qquad (1\text{-}32)$$

The rate constant k is a function only of temperature, so the integration of equation (1-32) gives directly:

$$C_A = C_{A_0} - kt \qquad (1\text{-}33)$$

First Order. The *first-order* case can be represented by the same model reaction, $A \longrightarrow B$, with the rate law

$$r_A = \frac{dC_A}{dt} = -kC_A \qquad (1\text{-}34)$$

For the same initial conditions,

$$\ell n\left(\frac{C_A}{C_{A_0}}\right) = -kt \qquad (1\text{-}35)$$

or if one is dealing with the course of the reaction between two specified time limits, C_{A_1}, at t_1 and C_{A_2} at t_2:

$$\ell n\left(\frac{C_{A_2}}{C_{A_1}}\right) = -k(t_2 - t_1) \qquad (1\text{-}36)$$

Second Order. *Second-order* reactions can be either of two general types: $2A \longrightarrow C + D$ or $A + B \longrightarrow C + D$. In the first case,

$$r_A = \frac{dC_A}{dt} = -kC_A^2 \qquad (1\text{-}37)$$

and

$$\frac{1}{C_A} - \frac{1}{C_{A_0}} = kt \qquad (1\text{-}38)$$

In the second case,

$$\frac{dC_A}{dt} = -kC_A C_B \qquad (1\text{-}39)$$

For this system values for both C_A and C_B must be specified for some value of time, say $t = 0$. However, if we look to the integration of equation (1-39),

$$\int_{C_{A_0}}^{C_A} \frac{dC_A}{C_A C_B} = -\int_0^t kt \qquad (1\text{-}40)$$

It is apparent that some relationship between C_A and C_B must be established before the integration can be carried out. This relationship is obtained from a combination of the specified values of C_A and C_B for a given time and the stoichiometry of the reaction. The stoichiometry of the reaction indicates that equimolal quantities of A and B react, so that if the reaction system is constant volume,

$$C_{A_0} - C_A = C_{B_0} - C_B \tag{1-41}$$

or

$$C_B = C_{B_0} - C_{A_0} + C_A$$

Then equation (1-40) becomes

$$\int_{C_{A_0}}^{C_A} \frac{dC_A}{C_A[(C_{B_0} - C_{A_0}) + C_A]} = -\int_0^t k\, dt \tag{1-40a}$$

and

$$\ln\left(\frac{C_A}{C_B}\right) - \ln\left(\frac{C_{A_0}}{C_{B_0}}\right) = -(C_{B_0} - C_{A_0})kt \tag{1-42}$$

When the reactants A and B are initially present in stoichiometric proportion, that is, $C_{A_0} = C_{B_0}$, equation (1-42) becomes indeterminate. The result, however, is given by equation (1-38), since the reaction stoichiometry requires that $C_A = C_B$ for $C_{A_0} = C_{B_0}$.

Nonintegral Order. A *nonintegral-order* rate equation, such as that for acetaldehyde decomposition [reaction (1) of Table 1.1a], does not fit into the pattern expected for the rates of true elementary steps, but it is convenient to consider it here in succession with the other simple order laws. Here we have, for example, A $\longrightarrow$ B, where the reaction is n-order with respect to A:

$$r_A = \frac{dC_A}{dt} = -kC_A^n \tag{1-43}$$

The integrated result for C_A is

$$\left(\frac{C_A}{C_{A_0}}\right)^{1-n} - 1 = \frac{n-1}{(C_{A_0})^{1-n}}kt \tag{1-44}$$

Half-Life, Conversion, and Volume Expansion Forms. At this point we are going to break into the narrative of rate forms to discuss three particular aspects of kinetic formulations which in general will pertain to most of the situations we discuss, at least for simple reactions.

The first of these are the concepts of *relaxation time* and reaction *half-life*. The rate equations we have written incorporate the constant k to express the proportionality between the rate and the state variables of the system, and as such it is some measure of a characteristic time constant for the reaction. For first-order reactions, in particular, a convenient association may be made, owing to the logarithmic time dependence of reactant concentration. If we rewrite equation (1-35) in exponential form, we see that the e- folding time of (C_A/C_{A_0}), that is, the time at which $-kt = -1$ and where (C_A/C_{A_0}) has decreased to 36.8% of its original value, occurs for $t = 1/k$. This value of time is sometimes referred to for first-order process as the relaxation time, τ, and gives directly the inverse of the rate constant. Since rate laws other than first-order do not obey exponential relations, this interpretation of relaxation time is strictly correct only for first-order systems; however, a related concept, that of reaction half-life, can be used more generally. As indicated by the name, reaction half-life specifies the time required for reaction of half the original reactant. These are also functions of the rate constant and are directly obtained from the integrated equations where $C_A = \frac{1}{2}C_{A_0}$ and $t = t_{1/2}$. For the cases of the previous section:

ZERO ORDER: $t_{1/2} = \dfrac{C_{A_0}}{2k}$

FIRST ORDER: $t_{1/2} = \dfrac{\ell n\, 2}{k}$

SECOND ORDER $(2A \longrightarrow C + D)$: $t_{1/2} = \dfrac{1}{kC_{A_0}}$

SECOND ORDER $(A + B \longrightarrow C + D)$: $t_{1/2} = \dfrac{1}{k(C_{A_0} - C_{B_0})}\ell n\left(\dfrac{C_{B_0}}{2C_{B_0} - C_{A_0}}\right)$

NONINTEGRAL ORDER: $t_{1/2} = \dfrac{[(\frac{1}{2})^{1-n} - 1]C_{A_0}^{1-n}}{(n-1)k}$

$$(1\text{-}45)$$

The second of these aspects has to do with whether we wish to write kinetic expressions from the point of view of reactant remaining, as we have done, or from the point of view of the product produced (reactant reacted). It is convenient in many instances to talk about the *conversion* in a reaction system, which defines the amount of reactant consumed or product made. For constant-volume systems this can be represented in the rate equations as a concentration of material reacted, C_x, so that $C_A = C_{A_0} - C_x$. For a first-order example, then

$$\frac{dC_A}{dt} = \frac{d(C_{A_0} - C_x)}{dt} = -k(C_{A_0} - C_x) \qquad (1\text{-}46)$$

and

$$\ell n\left(1 - \frac{C_x}{C_{A_0}}\right) = -kt \qquad (1\text{-}47)$$

corresponding to equations (1-34) and (1-35). The quantity (C_x/C_{A_0}) is a fraction varying between zero (no conversion) and 1 (complete conversion) for this irreversible reaction, and is ordinarily called the *fractional conversion of reactant*, x. This fractional-conversion definition in terms of concentration ratios applies only for constant-volume reaction systems, so we will define the fractional conversion of reactant more generally as the ratio of the moles reacted to initial moles present. Of course, corresponding definitions will pertain to mass amounts as well. Fractional conversions to product (moles product made to moles reactant consumed) are sometimes convenient to use for some of the complex reaction schemes we shall discuss later, but for simple reactions there is no particular advantage in using these terms. Corresponding expressions to equations (1-45) are:

ZERO ORDER: $\quad C_x = kt; \quad x = \dfrac{kt}{C_{A_0}}$

FIRST ORDER: $\quad \ell n \left(1 - \dfrac{C_x}{C_{A_0}}\right) = \ell n\,(1 - x) = -kt$

SECOND ORDER (2A $\longrightarrow$ C + D): $\dfrac{C_x}{C_{A_0}} = x = \dfrac{ktC_{A_0}}{1 + ktC_{A_0}}$

SECOND ORDER (A + B $\longrightarrow$ C + D): $\ell n \left(\dfrac{1 - C_x/C_{A_0}}{1 - C_x/C_{B_0}}\right) = -(C_{B_0} - C_{A_0})kt$

$$= \ell n \left(\dfrac{1 - x_A}{1 - x_B}\right)$$

NONINTEGRAL ORDER: $\left(1 - \dfrac{C_x}{C_{A_0}}\right)^{1-n} = 1 + \dfrac{n-1}{C_{A_0}^{1-n}} kt = (1 - x)^{1-n}$

$$(1-48)$$

　　The third of these aspects is how to modify the rate equations when the reaction-system volume is not a constant. Volume changes occur when there is a net change in the number of moles as the reaction proceeds. For most reactions that occur in the liquid phase such changes in volume are small and normally can be neglected. Conversely, in gas-phase reactions the volume/molal relation is governed in the limit by the ideal gas law and volume changes must be taken into account. For the simple reactions we have been discussing, it is possible to write the volume change directly in terms of the amount of reaction in a particularly convenient form, using the fractional conversion x. First define an expansion (or contraction) factor, ϵ, which represents the difference in reaction-system volume at the completion of reaction and at the start of reaction divided by the initial value:

$$\epsilon = \frac{V_\infty - V_0}{V_0} \qquad (1-49)$$

where V_0 is the initial volume and V_∞ the final volume. The relationship between the amount of reactant consumed and the reaction volume can now be determined from the stoichiometry of the particular reaction under consideration. As an example consider the irreversible first-order reaction A $\longrightarrow$ νB, where ν is some stoichiometric coefficient other than unity and is a positive quantity for products of reaction. When the volume/molal relation is a direct proportionality, as in gas-phase reactions,

$$\epsilon = \nu - 1 \qquad (1-50)$$

The number of moles of A at any time is

$$N_A = N_{A_0}(1 - x) \tag{1-51}$$

the volume of the reaction system is

$$V = V_0(1 + \epsilon x) \tag{1-52}$$

and the concentration of reactant A is

$$C_A = \frac{N_A}{V} = \frac{N_{A_0}(1 - x)}{V_0(1 + \epsilon x)} = C_{A_0}\frac{1 - x}{1 + \epsilon x} \tag{1-53}$$

We can now use these relationships in the basic reaction-rate definition. For the first-order irreversible batch reaction,

$$\frac{1}{V}\frac{dN_A}{dt} = -k\left(\frac{N_A}{V}\right) \tag{1-54}$$

and on substitution for V and N_A,

$$\frac{N_{A_0}}{V_0(1 + \epsilon x)}\frac{d(1 - x)}{dt} = -k\left(\frac{N_{A_0}}{V_0}\right)\frac{1 - x}{1 + \epsilon x}$$

or

$$\frac{1}{1 + \epsilon x}\frac{dx}{dt} = \frac{k(1 - x)}{1 + \epsilon x} \tag{1-55}$$

To integrate equation (1-55) we must keep in mind that the use of fractional conversions changes the limits of integration:

$$\int_0^x \frac{dx}{1 - x} = \int_0^t k\, dt \tag{1-56}$$

and

$$\ln(1 - x) = -kt \tag{1-57}$$

Interestingly, the result obtained here is the same as that for the constant-volume system, equation (1-48), since the expansion factor does not appear. This is a peculiarity only of first order reactions.[1] Keep in mind also that although fractional conversions are the same in the two systems, the concentrations will be different, as shown in equation (1-53). We can again give a listing, in terms of fractional conversion and expansion factors, of the equations for non-constant-volume systems.

ZERO ORDER: $C_{A_0}\ln(1 + \epsilon x) = \epsilon kt$

FIRST ORDER: $\ln(1 - x) = -kt$

SECOND ORDER $(2A \rightarrow \nu$ products): $\dfrac{(1 + \epsilon)x}{1 - x} + \epsilon \ln(1 - x) = C_{A_0}kt$ (1-58)

NONINTEGRAL ORDER: $\displaystyle\int_0^x \frac{(1 + \epsilon x)^{n-1}}{(1 - x)^n}\, dx = C_{A_0}^{n-1}kt$

---

[1] It is worthwhile to note that first-order reactions often do not follow the patterns expected from the behavior of non-first-order reactions. Another example is the first-order half-life, the only one independent of initial reactant concentration.

The integral in the last expression above is not a simple form and is most easily evaluated by graphical or numerical means. Use of the expansion factor is restricted to systems where there is a linear relationship between conversion and volume. For reactions that have complex sequences of steps, this linear relationship may not be true. Then we must rewrite the rate definition for a batch reactor:

$$r_A = \frac{1}{V}\frac{dN_A}{dt} = \frac{1}{V}\frac{d(C_A V)}{dt} = \frac{dC_A}{dt} + \frac{C_A}{V}\frac{dV}{dt} \tag{1-59}$$

and the volume terms must be evaluated from the detailed stoichiometry of the reaction steps.

c. Rates and conversions for some reversible reactions

The most important forms for reversible reactions are first-order, forward and reverse, and second-order, forward and reverse. If change in the number of moles occurs and order follows stoichiometry, then reversible reactions can also involve forward and reverse steps of different order. Since in the following we treat the reactions as elementary steps, the ratio of rate constants does define the equilibrium constant for the reaction; $K = k_f/k_r$.

First-Order Forward and Reverse. Here we have $A \rightleftharpoons B$:

$$r_A = \frac{dC_A}{dt} = -k_f C_A + k_r C_B$$

or

$$\frac{d(1-x)}{dt} = -k_f(1-x) + k_r\left(\frac{C_{B_0}}{C_{A_0}} + x\right) \tag{1-60}$$

where x is the conversion of A.

Equation (1-60) can be integrated to several possible forms, but a convenient one is

$$(k_f + k_r)t = \ln\left(\frac{\alpha}{\alpha - x}\right) \tag{1-61}$$

where

$$\alpha = \frac{k_f - (C_{B_0}/C_{A_0})k_r}{k_f + k_r}$$

Sometimes one encounters first-order reversible kinetics treated by a rate equation which is written in terms of concentrations relative to the equilibrium position of the system. If we define x_∞ as the conversion of A at equilibrium, then

$$\frac{dx}{dt} = k'[(1-x) - (1-x_\infty)] \tag{1-62}$$

and

$$\ln\left(\frac{x_\infty - x}{x_\infty}\right) = -k't \qquad (1\text{-}63)$$

The rate constant k' is equal to the sum of the forward and reverse rate constants, $k_f + k_r$, and one must exercise care in treating the temperature dependence of k', since it does not follow an Arrhenius law. The usefulness of equation (1-63) depends on the availability of data on equilibrium conversion.

Second-Order Forward and Reverse. The reaction is $A + B \rightleftharpoons C + D$ and

$$r_A = \frac{dC_A}{dt} = -k_f C_A C_B + k_r C_C C_D$$

or

$$C_{A_0}\frac{d(1-x)}{dt} = -k_f C_{A_0}(1-x)(C_{B_0} - C_{A_0}x) + k_r(C_{C_0} + C_{A_0}x)(C_{D_0} + C_{A_0}x)$$
$$(1\text{-}64)$$

The right side of equation (1-64) yields a quadratic function, which, with some manipulation, can be integrated to the following cumbersome form:

$$\ln\left[\frac{2\gamma x/(\beta - q^{1/2}) + (1/C_{A_0})}{2\gamma x/(\beta + q^{1/2}) + (1/C_{A_0})}\right] = q^{1/2}t \qquad (1\text{-}65)$$

where

$$\gamma = k_f - \frac{1}{K} \qquad q = \beta^2 - 4\alpha\gamma$$

$$\beta = -k_f\left[C_{A_0} + C_{B_0} + \frac{C_{C_0} + C_{D_0}}{K}\right]$$

$$\alpha = k_f\left[C_{A_0}C_{B_0} - \frac{1}{K}(C_{C_0}C_{D_0})\right]$$

$$K = \frac{k_f}{k_r}$$

Often C_{C_0} and C_{D_0} will be zero (no products in the reaction mixture initially) and the equations above can be simplified somewhat.

First-Order Forward, Second-Order Reverse. This scheme is often important in the kinetics of dissociation reactions in solution, $A \rightleftharpoons B + C$, where volume changes are negligible.

$$r_A = \frac{dC_A}{dt} = -k_f C_A + k_r C_B C_C$$
$$(1\text{-}66)$$
$$\frac{d(1-x)}{dt} = k_f(1-x) + C_{A_0}k_r x^2$$

where we have taken $C_{B_0} = C_{C_0} = 0$ for simplification. The conversion relationship is

$$2k_r\alpha t = \frac{1 + x(\beta - \tfrac{1}{2})}{1 - x(\beta + \tfrac{1}{2})} \qquad (1\text{-}67)$$

where

$$\alpha = \left[\frac{1}{4} \left(\frac{k_f}{k_r} \right)^2 + C_{A_0} \left(\frac{k_f}{k_r} \right) \right]^{1/2}$$

$$\beta = K\alpha$$

The integrated forms, equations (1-65) and (1-67), are as indicated awkward to work with and are particularly inconvenient to use in the interpretation of conversion data. If data on the equilibrium conversion, x_∞, are available, forms analogous to equation (1-63) may be written. These are, for the case of no products present initially:

$A + B \rightleftharpoons C + D$:

$$\ell n \left[\frac{(1/C_{A_0})(x_\infty + x) - 2x_\infty x}{(1/C_{A_0})(x_\infty - x)} \right] = 2k_f \left(\frac{1}{x_\infty} - 1 \right) C_{A_0} t \qquad (1\text{-}65a)$$

$A \rightleftharpoons B + C$:

$$\ell n \left[\frac{x_\infty(1 - x) + x}{x_\infty - x} \right] = k_f \left(\frac{2 - x_\infty}{x_\infty} \right) t \qquad (1\text{-}67a)$$

Equation (1-65a) is, in fact, one case of a general class of solutions for reversible reaction in terms of equilibrium conversion. If the right-hand side is written

$$\rho k_f \left(\frac{1}{x_\infty} - 1 \right) C_{A_0} t$$

then the following reactions are included in the solution:

Reaction	Value of ρ
$2A \rightleftharpoons B + C$	1
$A + B \rightleftharpoons C + D$	2
$2A \rightleftharpoons 2B$	2
$A + B \rightleftharpoons 2C$	4

d. Summary—simple reactions

The few examples given here provide only an indication of the variety of rate forms and conversion relationships which exist even for simple reactions. Certainly there are many more combinations of reaction order and reversibility that could be included, and some of these additional forms are involved in subsequent exercises and discussion without individual derivation. We will return to specifics of the manipulation of these rate and conversion expressions in later discussion of some methods for the interpretation of kinetic data and the influence of experimental error in determination of kinetic constants. While the formulation and integration of simple rate equations is not difficult, reflection on the nature of nonintegral and higher-order (> unity) rate laws indicates much more difficult mathematics, since these are nonlinear in the dependent concentration variables. Together with the exponential temperature dependence of rate, the nonlinear concentration dependence of rate in these cases means that even apparently simple reactions may be extremely difficult to analyze within the context of practical operation.

1.5   Kinetics of "Nearly Complex" Reaction Sequences

The fact that the kinetics of an overall reaction normally represent the net effect of the rates of a number of individual elementary steps means that one of our major concerns in analysis must be how to assemble the individual steps into the whole. There are systematic procedures for doing this, some of which will be discussed in detail for homogeneous reactions that occur by chain mechanisms. However, we have also pointed out that often, in the absence of detailed information, models that account for the essential features of the steps involved in an overall reaction can be of great utility. In this section we shall discuss three such schemes which have found application in a wide variety of reaction systems. These are:

Type I. Parallel reactions with separate reactants and products:

$$A \longrightarrow B \qquad (k_1)$$
$$L \longrightarrow M \qquad (k_2)$$

Type II. Parallel reactions with the same reactant:

$$A \longrightarrow B \qquad (k_1)$$
$$A \longrightarrow C \qquad (k_2)$$

Type III. Series reactions with a stable intermediate:

$$A \longrightarrow B \longrightarrow C \qquad (k_1, k_2)$$

a. Yield, selectivity, and kinetics

In addition to the rate of consumption of reactant or appearance of product, two further quantities are of importance in the analysis of the kinetics of these schemes: the *yield* and the *selectivity*. These quantities serve to specify the relative importance of the reaction paths that occur and are ordinarily based on some product to reactant or product to product relationship. Here we shall define yield as the fraction of reactant converted to a particular product, so if more than one product is of interest in a reaction, there can be several yields defined. The yield of product j with respect to reactant i, Y_j, is

$$Y_j = \frac{\text{moles of } i \text{ to } j}{\text{moles of } i \text{ present initially}}$$

Overall selectivity is defined by the following:

$$\text{yield} = (\text{overall selectivity})(\text{conversion of reactant}) \qquad (1\text{-}68)$$

Inserting the definitions for yield and conversion, we obtain

$$\text{overall selectivity} = \frac{\text{moles } i \text{ to } j}{\text{moles } i \text{ reacted}}$$

For some types of reaction schemes this overall selectivity is a function of the degree of conversion, so often it is convenient to define a differential or point selectivity referring to a particular conversion level:

$$\text{differential selectivity} = \frac{\text{rate of production of } j}{\text{rate of reaction of } i}$$

Application of these definitions to the three reaction sequences mentioned above will result in yield and selectivity values falling in the range from 0 to 1. If the stoichiometric coefficients of i and j differ, values outside this range may be obtained. In this case the definitions above can be normalized to the 0-to-1 range by multiplication by the ratio of the appropriate stoichiometric coefficients.

One should be careful in using these definitions in comparison with quantities given the same name elsewhere, since there is a considerable variation in the literature nomenclature. Further, the present definitions are dimensionless, while selectivities and yields reported in the literature, particularly those describing specific process results, may have mass or volume dimensions associated (i.e., grams product per liter of reactant, etc.).

Type I, II, and III schemes have been termed "nearly complex" here since they incorporate features of selectivity and yield associated with reaction systems consisting of a number of elementary steps, but certainly do not approach the level of detail required to describe completely such reactions. The analysis of rate, yield, and selectivity in such model systems is important in many chemical reaction engineering problems and is worth separate attention.

Type I. For parallel reactions with separate reactants, the analysis is easily handled, since the two steps are independent of each other. Indeed, the inclusion of Type I as a nearly complex scheme is not really necessary for the batch homogeneous reactions at constant volume treated here, since the yield and selectivity definitions are redundant with rate and conversion. The system is important in heterogeneous reactions, however, so we introduce it for the benefit of later reference. For first-order, irreversible reactions:

$$r_A = \frac{dC_A}{dt} = -k_1 C_A \qquad (1\text{-}69)$$

$$r_L = \frac{dC_L}{dt} = -k_2 C_L \qquad (1\text{-}70)$$

$$\ell n \left(\frac{C_A}{C_{A_0}}\right) = -k_1 t \tag{1-71}$$

and

$$\ell n \left(\frac{C_L}{C_{L_0}}\right) = -k_2 t \tag{1-72}$$

We may define a yield for each of the two reaction paths, but since there is no interaction between the two, the fraction reacted to a particular product is equal to the conversion for each. That is,

$$Y_B(I) = Y_M(I) = x$$

where $Y_B(I)$ and $Y_M(I)$ are the yields of B and M, respectively, in the Type I system. The differential selectivity is unity for each reaction and the overall selectivity, according to equation (1-68), must be unity. Unfortunately, yet another definition of selectivity creeps into the discussion at this point, since for the Type I system it is possible that the relative rates of the two reactions will be of interest in some applications—particularly true for some of the heterogeneous reactions we have mentioned. This we might term somewhat awkwardly a *parallel differential selectivity*, which would be given by

$$S_d(I) = \frac{-r_A}{-r_L} = \left(\frac{k_1}{k_2}\right)\left(\frac{C_A}{C_L}\right) \tag{1-73}$$

The rate constant ratio, (k_1/k_2), appears in equation (1-73), and this is given yet another name in some places, the *intrinsic selectivity*. The relationship between differential and intrinsic selectivity for the Type I system is shown in Figure 1.6.

Type II. For parallel reactions with the same reactant, the selectivity and yield definitions are more meaningful for homogeneous batch reactions. For the reactant the rates are additive, so

$$\frac{dC_A}{dt} = -(k_1 + k_2)C_A \tag{1-74}$$

and

$$\ell n \left(\frac{C_A}{C_{A_0}}\right) = -(k_1 + k_2)t \tag{1-75}$$

The yield of product B is

$$Y_B(II) = \frac{\text{reactant A converted to B}}{\text{initial reactant A}} = \frac{k_1}{k_1 + k_2}[1 - e^{-(k_1 + k_2)t}] \tag{1-76}$$

From equation (1-68), the overall selectivity is

$$S_o(II) = \frac{\text{yield}}{\text{conversion}} = \frac{k_1[1 - e^{-(k_1 + k_2)t}]}{(k_1 + k_2)[1 - e^{-(k_1 + k_2)t}]} \tag{1-77}$$

$$S_o(II) = \frac{k_1}{(k_1 + k_2)}$$

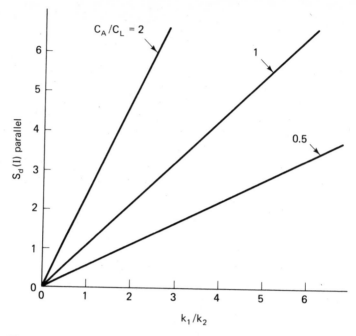

Figure 1.6 Relationship between differential and intrinsic selectivities for a Type I reaction system

The differential selectivity is given by the rate ratio[1]:

$$S_d(\text{II}) = \frac{\text{rate of production of B}}{\text{rate of consumption of A}} = \frac{k_1 C_A}{(k_1 + k_2)C_A}$$

$$S_d(\text{II}) = \frac{k_1}{k_1 + k_2} \tag{1-78}$$

which in this case is identical to the overall selectivity.

For Type II, the differential or overall selectivity and intrinsic selectivity are simply related by taking the inverse of equation (1-78) and rearranging:

$$S_i(\text{II}) = \frac{S_d(\text{II})}{1 - S_d(\text{II})} \tag{1-79}$$

This is shown in Figure 1.7. Note that the relationship is independent of concentration (conversion). Similar yield and selectivity definitions may be defined for product C.

---

[1]Some define differential selectivity as the ratio of a rate of production to a rate of reaction, in which case the denominator is a negative quantity and the corresponding selectivities negative numbers. We have used the definition in terms of rate of consumption of reactant only for the small convenience of working with positive quantities.

b. Other nearly complex reactions

Type I, II, and III reactions are not representative of all possible kinds of nearly complex sequences in which we may be interested. Series substitution reactions are often encountered, as in the case of formation of organic halides (problems 21 and 43), or the apparent kinetics of complicated reactions may be represented by higher-order series reactions somewhat analogous to Type III. Radioactive decay sequences are represented by an extended Type III sequence involving as many intermediates as unstable isotopes in the sequence. Extensive discussion of a number of these reactions are given by Benson (S. W. Benson, *The Foundations of Chemical Kinetics*, McGraw-

TABLE 1.1b
SOME EXAMPLES OF OTHER NEARLY COMPLEX REACTIONS

Reaction	Description or Example	Solution or Reference
$A \xrightarrow{k_1} C$ $B \xrightarrow{k_2} C$	Radioactive decay to common product	$\begin{cases} C_A = C_{A_0}e^{-k_1 t} \\ C_B = C_{B_0}e^{-k_2 t} \\ C_C = C_{A_0} + C_{B_0} - (C_A + C_B) \end{cases}$
$A \xrightarrow{k_1} B \xrightarrow{k_2} C \xrightarrow{k_3} D \rightarrow \dots$	Radioactive decay sequence	$C_i = a_{i1}e^{-k_1 t} + a_{i2}e^{-k_2 t}$ $+ \dots + a_{ii}e^{-k_i t}$ $a_{i1}, \dots =$ constants dependent on initial concentrations
$A \rightleftharpoons B \rightleftharpoons C \dots$	Series reversible reactions	Problem 20
$A + B \rightarrow C + D$ $C + B \rightarrow D + E$	Competitive second-order—series (Saponification of esters)	Problem 23
$A + B \rightarrow D + E$ $A + C \rightarrow F + E$	Competitive second-order—parallel (saponification of esters)	J. G. Van der Corput and H. S. Backer, *Proc. Acad. Sci. Amsterdam*, **41**, 1508 (1938)
$A + B \rightarrow C + D$ $C + B \rightarrow E + D$ $E + B \rightarrow F + D$	Series substitution	Problem 21
$A \rightarrow B$ $2B \rightarrow C$	Higher-order series reactions	J. Chien, *J. Amer. Chem. Soc.*, **70**, 2256 (1948)
$2A \rightarrow B \rightarrow C$	Higher-order series reactions	$\begin{cases} \text{Problem 22} \\ \text{V. W. Weekman, Jr., and D. M.} \\ \quad \text{Nace, } \textit{Amer. Inst. Chem. Eng.} \\ \quad \textit{J., } \textbf{16}, 397 (1970) \end{cases}$
$\begin{smallmatrix} & A & \\ \swarrow & & \searrow \\ C & \rightleftharpoons & B \end{smallmatrix}$	General series and parallel reactions	J. Wei and C. D. Prater, *Amer. Inst. Chem. Eng. J.*, **9**, 77 (1963)

Hill, New York, 1960, ch. III) and Frost and Pearson (A. A. Frost and R. G. Pearson, *Kinetics and Mechanism*, 2nd ed., Wiley, New York, 1961, ch. 8). Some examples are given in Table 1.1b. In addition to the solution for conversion shown, definitions of selectivity and yield analogous to those just discussed can be made for these reactions and will be of similar importance in definition of the relative efficiencies of the various reaction paths.

1.6 Kinetics of Complex Reaction Sequences— Chain Reactions

a. Individual steps and the pseudo-steady-state hypothesis

In our previous discussion of the elementary steps involved in chemical reactions we used the decomposition of diethyl ether as an example of a chain reaction in which a cycle of elementary steps produce the final products. There is a large number of reactions which are known to occur by chain mechanisms, and in the following we shall refer primarily to those which generally correspond to the closed sequence in the classification of Boudart. Here active centers (also called *active intermediates* or *chain carriers*) are reacted in one step and regenerated in another in the sequence; however, if we look back to reaction (IV) a closer examination discloses that some of the steps serve particular functions. In (IVa), the active centers $\cdot CH_3$ and $\cdot CH_2OC_2H_5$ are formed by the initial decomposition of the ether molecule, and in (IVd) they recombine to produce the ether. However, the overall products of the decomposition, C_2H_6 and CH_3CHO, are formed in the intermediate steps (IVb) and (IVc). In analysis of most chain reactions we can think of the sequence of steps as involving three principal processes:

1. *Initiation reactions*: reactions that provide the source of active intermediates in the chain.
2. *Propagation reactions*: reactions that involve the consumption and regeneration of active intermediates and in which stable products are formed.
3. *Termination or breaking reactions*: reactions that remove active intermediates from the chain (no regeneration).

The full analysis of the kinetics of a chain reaction such as this ether decomposition would require solution of the rate equations for each of the species involved, subject to the material-balance restrictions. Except in special cases this cannot be done analytically, and it has become customary to employ the *pseudo-steady-state hypothesis* (pssh) to simplify the mathematical problems involved in solving the coupled rate equations for chain mechanisms.

A commonly quoted version of this principle holds that since the concentrations of active centers are small, their rates of change are negligible throughout the course of the reaction. This is an unfortunate statement, for the one does not follow the other; rather, the success of the pssh depends on a small ratio of total active centers to initial reactant [F. G. Heineken, H. M. Tsuchiya, and R. Aris, *Math-Biosciences*, **1**, 95 (1967)]. A simple example of this is provided by a form of the Type III sequence we have just discussed. Consider that the overall transformation from A to C is promoted by a catalytic surface, in which B represents a surface intermediate. Thus,

$$A + S \longrightarrow B$$

$$B \longrightarrow C + S$$

where S is an empty site and both S and B are active centers. The rates of the elementary steps are

$$r_1 = k_1 C_A C_S \tag{1-93}$$

$$r_2 = k_2 C_B \tag{1-94}$$

and according to the pssh the rates of change of the active centers are zero; for a batch reactor we have

$$\frac{dC_S}{dt} = 0 = -k_1 C_A C_S + k_2 C_B \tag{1-95}$$

$$\frac{dC_B}{dt} = 0 = k_1 C_A C_S - k_2 C_B \tag{1-96}$$

Consistent with equations (1-95) and (1-96) is the constancy of total active-center concentration:

$$C_{S_0} = C_B + C_S \tag{1-97}$$

One of the pair (1-95) and (1-96) may be solved together with equation (1-97) to give the pssh concentration of the active center B:

$$C_B = \frac{C_{S_0}}{1 + (k_2/k_1)\dfrac{1}{C_A}} \tag{1-98}$$

The rate of formation of product is given by equation (1-94), so on substitution for C_B:

$$r_2 = \frac{k_2 C_{S_0} C_A}{(k_2/k_1) + C_A} \tag{1-99}$$

Now let us examine this system to see what conditions are necessary to make the pssh valid. Write the rate equations for A and B:

$$\frac{dC_A}{dt} = -r_1 = -k_1 C_A C_S \tag{1-100}$$

$$\frac{dC_B}{dt} = r_1 - r_2 = k_1 C_A C_S - k_2 C_B \tag{1-101}$$

with the initial conditions

$$t = 0, \quad C_A = C_{A_0}, \quad C_S = C_{S_0}, \quad C_B = 0$$

Now let us define the following dimensionless variables:

$$y = \frac{C_A}{C_{A_0}}, \quad z = \frac{C_S}{C_{S_0}}, \quad \mu = \frac{C_{S_0}}{C_{A_0}}$$

$$\tau = k_1 C_{S_0} t \quad \lambda = \frac{k_2}{k_1 C_{A_0}}$$

(1-102)

When these definitions are substituted into equations (1-100) and (1-101) the following are obtained:

$$\frac{dy}{d\tau} = -yz \tag{1-103}$$

$$\mu \frac{dz}{d\tau} = -yz + (1 - z)\lambda \tag{1-104}$$

The quantity μ is the key to understanding the problem. For $\mu = 0$ the second equation above reduces to the pssh form for C_S. The success of the pssh requires that the value of μ be small, not that the derivative $(dz/d\tau)$ approach zero. Such a condition is ensured by maintaining a small ratio of total active centers to the reactant concentration. If a chain reaction is to be efficient, this condition should be approximated in the sense that relatively few active centers would be required to produce many molecules of product in the propagation reactions. Although the illustration here is for a heterogeneous reaction, the same principle applies to homogeneous reactions.

b. General analysis of kinetics

We may now apply the pssh to a general treatment of chain reactions. The following development is an adaptation of the excellent discussion given by Frost and Pearson. Let us write the three parts of the chain reaction as follows:

INITIATION: $\ldots \ldots \longrightarrow m\text{R} + \ldots$

PROPAGATION: $\text{R} + \ldots \longrightarrow \text{R} + \ldots$ (IX)

TERMINATION: $\text{R} + \text{R} \ldots \longrightarrow$

In this sequence R represents the active center, which normally is an atom or radical in homogeneous reactions and a catalytic site in heterogeneous reactions. Other molecules involved in each step are not shown explicitly, and we let their concentration product with the appropriate rate constants of unit stoichiometric coefficient for R be designated by r_i for the initiation step, r_p for the propagation step, and r_t for the termination step. The rate equation for the active center R is

$$\frac{dR}{dt} = mr_i - Rr_p + Rr_p - 2r_tR^2 \tag{1-105}$$

If the pssh may be applied, equation (1-105) becomes

$$mr_i - 2r_tR^2 = 0 \tag{1-106}$$

or

$$R = \left(\frac{mr_i}{2r_t}\right)^{1/2} \tag{1-107}$$

Now if the formation of product in the chain occurs in the propagation step, the rate of product accumulation is

$$\text{propagation rate} = Rr_p = r_p\left(\frac{mr_i}{2r_t}\right)^{1/2} \tag{1-108}$$

The rates of the other individual steps are:

$$\text{initiation rate} = r_i$$

$$\text{termination rate} = r_tR^2 = \frac{mr_i}{2}$$

It is convenient in the classification of general schemes such as (IX) to characterize the chain both by the number of active centers involved and by the kinetic order of the chain termination step. Thus, (IX) would be referred to as involving a single active center with second-order termination. The ether decomposition reaction we have discussed previously is a chain with two active centers and second-order termination, first-order with respect to each of the active centers. In general notation:

INITIATION:	$\ldots \longrightarrow mR + nS + \ldots$
PROPAGATION:	$R + \ldots \longrightarrow S + \ldots$
	$S + \ldots \longrightarrow R + \ldots$
TERMINATION:	$R + S \longrightarrow \ldots$
INITIATION RATE:	mr_i for R
	nr_i for S
PROPAGATION RATE:	Rr_{p_1} for the first step
	Sr_{p_2} for the second step
TERMINATION RATE:	RSr_t

(X)

Here we apply the pssh to both active centers R and S.

$$mr_i - Rr_{p_1} + Sr_{p_2} - RSr_t = 0 \tag{1-109}$$

$$nr_i + Rr_{p_1} - Sr_{p_2} - RSr_t = 0 \tag{1-110}$$

These may be solved for R and S and the result substituted into the propagation-rate expression. For the first step we have propagation step 1 rate $= Rr_{p_1}$

$$= \left(\frac{m-n}{4}\right)r_i + \sqrt{\left[\left(\frac{m-n}{4}\right)r_i\right]^2 + \frac{(m+n)r_i r_{p_1} r_{p_2}}{2r_t}}$$

(1-111)

This is a cumbersome expression and really of little use to the analyst as it stands. Once again, however, we may make use of information concerning the relative magnitude of quantities which are involved in the rate equations, much in the way that it was possible to verify the pssh. In this case one looks to the relative magnitudes of the initiation rate, r_i, and the propagation rate, r_p. If the chain reaction is an efficient one (which it must be in order to occur by a chain mechanism), $r_p \gg r_i$. Then all terms that contain only r_i are negligible compared to others, and equation (1-111) becomes

$$Rr_{p_1} = \left[\frac{(m+n)r_i r_{p_1} r_{p_2}}{2r_t}\right]^{1/2}$$

(1-112)

The efficiency of a chain mechanism is often discussed in terms of the *chain length*. A number of different quantitative definitions have been used for chain length, but all give some measure of the number of chain-propagation (product-yielding) steps resulting from a single active center. A reaction with long chain length thus produces many chain-propagation steps per active center and is correspondingly an efficient one. The analysis used to obtain equation (1-112) from (1-111) is often termed the *long-chain approximation*. We will define chain length, v, as the ratio of the rate of a particular propagation reaction to the rate of formation of chain carriers involved in the propagation step. For the single chain carrier with second-order termination, reaction (IX), we obtain

$$v = \frac{r_p}{(2mr_i r_t)^{1/2}}$$

(1-113)

and for the first propagation step of (X),

$$v_1 = \frac{r_{p_1}}{[2(m+n)r_i r_t]^{1/2}}$$

(1-114)

It is apparent that the analysis developed for the chain mechanisms illustrated in (IX) and (X) may be extended to various other types of chains. In Table 1.2 is given a summary of results for some of the more common types of chain mechanisms, together with examples of reactions that are known to conform to them.

The specific propagation-rate equations derived for examples A–D in Table 1.2 are special cases of the general proportionality given under E. It is interesting to note that the order of the chain termination step is important in determining the form of the overall kinetic expression, and in this sense the termination reaction is more important than the initiation in its influence

TABLE 1.2
KINETICS OF SOME CHAIN REACTION MECHANISMS

A. Single Active Center with First-Order Termination

Chain:

$$\ldots \longrightarrow m\text{R} + \ldots$$

$$\text{R} + \ldots \longrightarrow \text{R} + \ldots$$

$$\text{R} + \text{M} + \ldots \longrightarrow \ldots$$

Propagation Rate:

$$r_p R = r_p\left(\frac{mr_i}{r_t}\right)$$

Example:

First-order termination steps are often associated with extraneous factors, for example the adsorption of chain carriers on the walls of a reaction vessel.

B. Single Active Center with Second-Order Termination

Chain:

$$\ldots \longrightarrow m\text{R} + \ldots$$

$$\text{R} + \ldots \longrightarrow \text{R} + \ldots$$

$$\text{R} + \text{R} \ldots \longrightarrow \ldots$$

Propagation Rate:

$$r_p R = r_p\left(\frac{mr_i}{2r_t}\right)^{1/2}$$

Example: Ortho-Para Hydrogen Conversion

$$\text{M} + \text{H}_2(o \text{ or } p) \longrightarrow 2\text{H} + \text{M} \qquad (k_i)$$

$$\text{H} + \text{H}_2(p) \longrightarrow \text{H}_2(o) + \text{H} \qquad (k_p)$$

$$\text{M} + 2\text{H} \longrightarrow \text{H}_2(o \text{ or } p) + \text{M} \qquad (k_t)$$

$$R = [\text{H}]$$

$$r_p R = k_p\left(\frac{k_i}{k_t}\right)^{1/2}[\text{H}_2]^{1/2}[\text{H}_2(p)]$$

C. Two Active Centers with Second-Order Termination

Chain:

$$\ldots \longrightarrow m\text{R} + n\text{S} + \ldots$$

$$\text{R} + \ldots \longrightarrow \text{S} + \ldots$$

$$\text{S} + \ldots \longrightarrow \text{R} + \ldots$$

$$2\text{R} + \ldots \longrightarrow \ldots$$

Propagation Rate (Long-Chain Approximation):

$$r_{p_1} R = r_{p_1}\left[\frac{(m + n)r_i}{2r_t}\right]^{1/2}$$

$$r_{p_2} S = r_{p_1}\left[\frac{(m + n)r_i}{2r_t}\right]^{1/2}$$

TABLE 1.2 (Cont.)

Example: $H_2 + Br_2 \rightleftharpoons 2HBr$

$$Br_2 \longrightarrow 2Br \qquad (k_i)$$

$$Br + H_2 \longrightarrow HBr + H \qquad (k_1)\}Rr_{p_1}$$

$$\left.\begin{array}{l} H + Br_2 \longrightarrow HBr + Br \qquad (k_2) \\ H + HBr \longrightarrow H_2 + Br \qquad (k_{-1}) \end{array}\right\} Sr_{p_2} = \dfrac{k_2[H][Br_2] - k_1[HBr][H]}{k_2[H][Br_2] + k_{-1}[H][HBr]}$$

$$2Br \longrightarrow Br_2 \qquad (k_t)$$

$$R = [Br], \quad S = [H], \quad m = 2, \quad n = 0$$

$$r_{p_1}R + r_{p_2}S = \frac{2k_1(k_i/k_t)^{1/2}[H_2][Br_2]^{1/2}}{1 + (k_{-1}[HBr]/k_2[Br_2])}$$

D. Two Active Centers with Second-Order Cross-Termination

Chain:

$$\cdots \longrightarrow mR + nS + \cdots$$
$$R + \cdots \longrightarrow S + \cdots$$
$$S + \cdots \longrightarrow R + \cdots$$
$$R + S + \cdots \longrightarrow \cdots$$

Propagation Rate (Long-Chain Approximation):

$$r_{p_1}R = \left[\frac{(m+n)r_i r_{p_1} r_{p_2}}{2r_t}\right]^{1/2}$$

$$r_{p_2}S = \left[\frac{(m+n)r_i r_{p_1} r_{p_2}}{2r_t}\right]^{1/2}$$

Example: $C_2H_5OC_2H_5 \longrightarrow C_2H_6 + CH_3CHO$

$$C_2H_5OC_2H_5 \longrightarrow \cdot CH_3 + \cdot CH_2OC_2H_5 \qquad (k_i)$$
$$\cdot CH_3 + C_2H_5OC_2H_5 \longrightarrow C_2H_6 + \cdot CH_2OC_2H_5 \qquad (k_1)$$
$$\cdot CH_2OC_2H_5 \longrightarrow \cdot CH_3 + CH_3CHO \qquad (k_2)$$
$$\cdot CH_3 + \cdot CH_2OC_2H_5 \longrightarrow C_2H_5OC_2H_5 \qquad (k_t)$$

$$R = [\cdot CH_3], \quad S = [\cdot CH_2OC_2H_5], \quad m = 1, \quad n = 1$$

$$r_{p_1}R + r_{p_2}S = 2\left(\frac{k_i k_1 k_2}{k_t}\right)^{1/2}[C_2H_5OC_2H_5]$$

E. Some Generalized Results

1. Rate of propagation $\propto r_p \left(\dfrac{r_i}{r_t}\right)^{1/w}$

where w = order of chain termination step with respect to active center

2. Chain length $\propto \left(\dfrac{r_p}{r_i}\right)\left(\dfrac{r_i}{r_t}\right)^{1/w}$

on the behavior of the chain. The reader may verify the general proportionality given in E for chain length.

c. Some special types of chain reactions
and their analysis

Since chain reactions are so common, it would be unreasonable to expect all mechanisms to fit neatly into the four examples of Table 1.2. Before extending the discussion, however, we should add a few comments about the two reactions selected as illustrative of two-active-center propagation steps.

The first of these, the reaction of hydrogen and bromine vapor, is probably the first reaction for which a suitable chain mechanism was identified. The kinetics of this reaction were carefully studied and reported as early as 1907 [M. Bodenstein and S. C. Lind, *Z. Physik. Chem.*, **57**, 168 (1907)] and the chain-reaction-mechanism interpretation of the reported kinetics, as shown in Table 1.2, given over a decade later [J. A. Christiansen, *Kgl. Danske Videnskab. Selskab.*, **1**, 14 (1919); K. F. Herzfeld, *Ann. Physik*, **59**, 635 (1919); M. Polanyi, *Z. Elektrochem.*, **26**, 50 (1920)]. A thorough discussion of this reaction is given in the text by Frost and Pearson.

The second two-active-center example is characteristic of the mechanism of thermal decomposition of many organic molecules. The propagation rate is seen to be first order in the decomposing ether, giving the false impression of a unimolecular reaction. This particular scheme is known as a *Rice–Herzfeld mechanism* after the workers [F. O. Rice and K. F. Herzfeld, *J. Amer. Chem. Soc.*, **56**, 284 (1934)] who proposed chain mechanisms with the cross-termination step for decomposition reactions and showed how simple first-order kinetics could pertain to this type of chain. Cross termination, however, is not a uniform property of organic thermal decompositions; the pyrolysis of acetaldehyde, which has been studied extensively and which is described by a Rice–Herzfeld mechanism with $\cdot CH_3$ and $\cdot CH_3CO$ as the active centers, involves the recombination of methyl radicals as the termination step. In accord with the general proportionality of E in Table 1.2, the kinetics of this reaction are three-halves-order with respect to the acetaldehyde.

A very important type of chain mechanism, particularly in polymerization processes, is the *chain transfer* reaction, in which an active center reacts with a molecule to form a new active center not involved in the original chain. Addition polymerization, such as the formation of polystyrene, is a good example. In these reactions one uses an initiator, usually a molecule that is easily decomposed thermally, to produce a radical which then reacts with the organic molecule, producing a new radical that becomes both the effective active center and the growing product molecule. In this case:

$$\ldots \longrightarrow 2R$$
$$R + M \longrightarrow R'$$
$$R' + M \longrightarrow R''$$
$$2R'' \longrightarrow \ldots$$

(XI)

The active center R produced in the initiation step adds to the monomer molecule to produce the new active center, R', which in turn adds to another monomer molecule, giving R''; and so on. The termination step involves a disproportionation or recombination of two active centers giving the final polymer product. Obviously, there will be a distribution of active centers (i.e., polymer molecules that have been growing for different lengths of time), so the polymer product we are discussing is not a uniform one but will consist of a range of molecules of different sizes. In practice, one can control many of the properties of the polymer product by controlling this distribution of size or molecular weight. The initiator, which chemically may be quite different from the substance being polymerized, constitutes only a negligible portion of the final polymer molecule.

The kinetics of chain transfer reactions are generally first-order in monomer concentration and half-order in the initiator. This result is consistent with the chain steps of (XI) if the propagation rate constant is the same regardless of the size of the active center; this is true, for example, in styrene polymerization, where k_p is about 10^3 (moles-sec)$^{-1}$ at 80°C.

A final, interesting type of chain mechanism is the branching chain reaction. In this case one of the chain steps (branching) yields two active centers from the reaction of one:

$$\ldots \longrightarrow m\text{R}$$
$$\text{R} + \ldots \longrightarrow \text{R} + \ldots$$
$$\text{R} + \ldots \longrightarrow 2\text{R} + \ldots \qquad \text{(XII)}$$
$$\text{R} + \ldots \longrightarrow$$

The pssh gives for the concentration of R:

$$\text{R} = \frac{mr_i}{r_t - r_b} \qquad (1\text{-}115)$$

where r_b is defined for the branching step in the same manner as r_i and r_t for initiation and termination. The propagation rate is

$$r_p\text{R} = \frac{mr_i r_p}{r_t - r_b} \qquad (1\text{-}116)$$

which has the interesting property of approaching infinity as the magnitude of r_t and r_b become the same. Physically this means an explosion, so $r_t = r_b$ defines an explosion limit for the branching chain reaction. The branching chain theory has been used to interpret explosions in the low-pressure region for reactions such as H_2-O_2 and $CO-O_2$. One should keep in mind that this sort of explosion is different from the normal thermal explosion in which the heat evolved in a reaction cannot be removed from the system, leading to increasing temperatures, rates of reaction, and heat evolution.

d. Efficiency of chain mechanisms

The reason that so many chemical reactions proceed by a chain of individual steps involving active centers, rather than occurring directly, obviously must be due to the fact that the chain mechanisms are more efficient in promoting the transformation. This efficiency can be translated directly into the energy requirement of the reaction; chain mechanisms (if they occur) have lower energy requirements than alternative direct reactions. Let us take for the moment the activation energy as a measure of the energy requirements of a reaction, and write the generalized rates r_i, r_p and r_t in terms of their temperature dependence

$$r_i = r_i^\circ e^{-E_i/RT}$$

$$r_p = r_p^\circ e^{-E_p/RT} \qquad\qquad (1\text{-}117)$$

$$r_t = r_t^\circ e^{-E_t/RT}$$

Now substitute these forms into the general expression for propagation rate from E in Table 1.2:

$$\text{rate of propagation} \propto r_p^\circ e^{-E_p/RT}\left(\frac{r_i^\circ}{r_t^\circ}\frac{e^{-E_i/RT}}{e^{-E_t/RT}}\right)^{1/w}$$

or

$$\text{rate of propagation} \propto e^{-[E_p+(1/w)(E_i-E_t)]/RT} \qquad\qquad (1\text{-}118)$$

The apparent activation energy of the overall chain reaction is thus not the same as the activation energy for the propagation step, nor is it a direct function of the activation energy for the initiation step. In fact, the apparent activation energy can be considerably smaller than E_i, due either to $E_t > 0$ or second-order termination where $w = 2$. For this reason, a chemical transformation may occur by a chain mechanism even if the chain involves an initiation step with higher activation energy than the alternative direct reaction.

The discussion above may shed some light on *why* chain reactions occur, but the thoughtful reader will doubtless have asked himself by now how one determines *which* reactions are involved in a given chain. The answer is not simple, since reproduction of the form of a rate expression determined experimentally cannot be taken as verification of a proposed chain mechanism. It is necessary to examine other plausible reaction steps and to show that their rates are negligibly slow compared to those in the proposed mechanism. This is a complicated task but can be accomplished by careful consideration of the activation energies of alternative reactions, of the bond energies involved in these reactions, and of the concentration levels of the various active centers appearing in the chain. These factors will be considered in detail in chapter 2.

1.7 Reaction Rates and Conversion in Nonisothermal Systems

Thus far in discussion kinetics we have confined our interests to the analysis of expressions of the form

$$\int f(C_A, C_B, \ldots)dC_A = \int k \, dt \qquad (1\text{-}119)$$

for the example of a constant-volume batch system, where $f(C_A, C_B, \ldots)$ denotes the form of the concentration dependence of rate. In nonisothermal systems the temperature dependence of the rate, which is far more pronounced than its concentration dependence, becomes important and, on substitution of the Arrhenius form for k in equation (1-119), we end up with a much more formidable problem to solve

$$\int f(C_A, C_B, \ldots)dC_A = \int k^{\circ} e^{-E/RT} \, dt \qquad (1\text{-}120)$$

From one point of view, the solution depends on obtaining a relationship between temperature and composition, so the terms of equation (1-120) can be expressed in terms of a single dependent variable and, hopefully, integrated. Alternatively, one could look for the temperature/time relationship. This may not be so easy to do, however, since the time/temperature or composition/temperature history of a reaction in which heat is evolved or consumed is a function of the rate itself. Obviously, one must look to another relationship in addition to that of mass conservation in order to obtain this history.

a. Coupling of the energy-conservation equation to the mass-conservation equation

In isothermal systems, which we have considered thus far, the general mass conservation/reaction rate expression of equation (1-15) is sufficient to describe the state of the system at any time. In nonisothermal systems this is not so, and expressions both for the conservation of mass and the conservation of energy are required. In reacting systems the energy balance most conveniently is written in terms of an enthalpy balance for all the species entering and leaving a reference volume such as that of Figure 1.3a. Chemical reaction affects this balance by the heat that is evolved or consumed in the reaction. The enthalpy balance that is required in addition to equation (1-15) to describe a nonisothermal system is

$$\left(\frac{\text{enthalpy entering } \delta}{\text{time}}\right) - \left(\frac{\text{enthalpy leaving } \delta}{\text{time}}\right) + \left(\frac{\text{enthalpy produced in } \delta}{\text{time}}\right)$$

$$= \left(\frac{\text{change of enthalpy in } \delta}{\text{time}}\right)$$

$$(1\text{-}121)$$

$$[(N_{A_0} - N)C_{P_A} + (N_{B_0} - N)C_{P_B} + NC_{P_C}](T - T_0) - N(-\Delta H_R) = 0 \quad (1\text{-}135)$$

Combining equations (1-133) and (1-132a) and integrating, we obtain

$$t = \int_0^{N_{A_0}, N_{B_0}} \frac{dN}{\dfrac{k_1^\circ e^{-E_1/RT}(N_{A_0} - N)(N_{B_0} - N)}{N_{A_0} + N_{B_0} - N}\left(\dfrac{P}{RT}\right) - k_2^\circ e^{-E_2/RT} N} \quad (1\text{-}136)$$

where the upper limit refers to which of the two reactants is initially present in the smallest quantity (limiting reactant). Equations (1-135) and (1-136) can be solved simultaneously by the following iterative procedure:

1. Assume a value for the temperature. If the reaction is exothermic, $T > T_0$.
2. Knowing the value of T_0 and T, together with N_{A_0}, N_{B_0}, and the heat capacities, calculate the corresponding N from equation (1-135).
3. Calculate the corresponding value of t by graphical or numerical integration of equation (1-136).

As a result of the difficulties involved in obtaining general analytical solutions for these nonisothermal problems, one sometimes encounters approximations which are based on simpler forms of the temperature dependence of the reaction-rate constant. The two most commonly used are

$$k = a + bT \quad (1\text{-}137)$$

$$k = a^\circ\left(\frac{T}{T_0}\right)^n \quad (1\text{-}138)$$

The dangers inherent in the linear approximation to an exponential function should be obvious and equation (1-137) is useful only when small changes in temperature are anticipated. Equation (1-138) is a more flexible approximation; however, in many cases it yields solutions in the form of series expansions which are ultimately no more convenient numerically than the simultaneous solution of equations (1-135) and (1-136). A comparison of these approximation methods for an example with activation energy of 20,000 cal/mole and constant of unity at 300°K is shown in Figure 1.10. The fit has been forced at 300°K and 320°K in this illustration. For the linear approximation there are large deviations even within the range of the fit; the power-law form is much better but only in the range of the fit.

c. Temperature effects in some nonelementary
reaction sequences

We have seen that one important reason for the efficiency of chain reaction mechanisms is the manner in which the combination of activation energies for initiation, propagation, and termination steps, equation (1-118), gives a net energy requirement less than that for the initiation step. In this

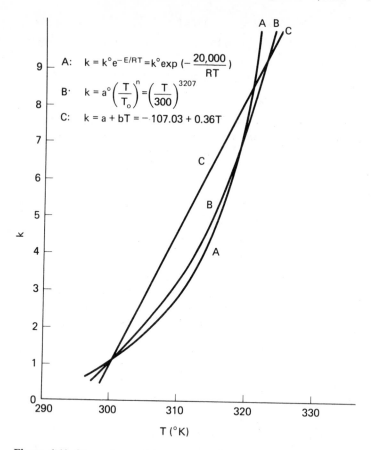

Figure 1.10 Comparison of temperature dependence of rate-constant approximations: $E = 20{,}000$ cal/mole, $k = 1$ at $300°K$

sense, we should talk a little about the influence of temperature on the behavior of the examples of nearly complex reactions discussed previously. We shall not attempt solution of these systems for adiabatic conditions here, but rather examine the results for conversion and selectivity in terms of how they are affected by temperature level.

Rewriting the differential selectivity definitions made previously for Types I, II, and III in terms of the temperature dependence:

$$S_d(\text{I}) = \left(\frac{k_1^\circ}{k_2^\circ}\right) \exp\left(\frac{E_2 - E_1}{RT}\right)\frac{C_A}{C_L} \tag{1-139}$$

$$S_d(\text{II}) = \frac{1}{1 + \left(\frac{k_2^\circ}{k_1^\circ}\right) \exp\left(\frac{E_1 - E_2}{RT}\right)} \tag{1-140}$$

$$S_d(\text{III}) = 1 - \left(\frac{k_2^\circ}{k_1^\circ}\right) \exp\left(\frac{E_1 - E_2}{RT}\right)\frac{C_B}{C_A} \tag{1-141}$$

In each of these cases, the differential selectivity will increase as the reaction temperature is increased if $E_1 > E_2$, and will decrease if $E_2 > E_1$. Of course, trends in selectivity with temperature are not always the same as for these three examples; it is entirely possible for selectivity to decrease with increasing temperature for some systems (see problem 30).

A corresponding analysis may be applied to yields and overall selectivity. Since the activation energy is conferred upon a reaction by nature, there is little one can do to change the trends of these measures of efficiency with temperature; however, if such information is available, it can be used to the designer's favor. An example of this is given in the following section.

d. Temperature-scheduled reactions

At the beginning of this section on nonisothermal reactions we introduced the idea that the solution to the kinetic problem could be viewed as one of obtaining time/temperature or composition/temperature histories for the reaction. For adiabatic reactions we found how these relationships could be determined from the mass and enthalpy conservation equations. The same principles would apply for nonadiabatic systems provided that we can specify the details of heat transfer to the surroundings.

Let us now consider such a nonadiabatic problem. We still will not worry about the details of heat transfer, but will assume that we can maintain by some means any desired level of temperature at any instant in the reaction system. The question is what, if any, advantages might one be able to obtain by using some schedule of temperature, varying with time of reaction. For what kinds of reactions might such procedures be valuable?

As it turns out, there is a very important class of reactions, those which are equilibrium-limited, for which temperature scheduling can be quite beneficial. We will take the simple example [O. Bilous and N. R. Amundson, *Chem. Eng. Sci.*, **5**, 81 (1956)] of a reversible, first-order reaction:

$$A \rightleftharpoons B \qquad (k_f, k_r)$$

for which at low temperatures the equilibrium lies far to the right but the rate of reaction of A is slow, and for which at higher temperatures the rate of both forward and reverse reactions is large ($E_r > E_f$). What sort of temperature schedule can we impose on this reaction to obtain a given degree of conversion to B in the absolute minimum time? For a constant-volume reaction the rate equation is

$$\frac{dC_A}{dt} = -k_f C_A + k_r C_B \qquad (1\text{-}60)$$

and the time of reaction is

$$t = \int_0^x \frac{C_{A_0}\, dx}{k_f C_{A_0} - k_r C_{B_0} - (k_f + k_r) C_{A_0} x} \qquad (1\text{-}142)$$

The problem is to define the path of x versus time which will minimize this integral. For this reaction system, minimization of the integral corresponds to maximization of the rate of conversion to B at each instant, or

$$\frac{\partial}{\partial T}\left[k_f C_{A_0} - k_r C_{B_0} - (k_f + k_r)x\right] = 0 \qquad (1\text{-}143)$$

This reduces to

$$C_{A_0}\frac{\partial k_f}{\partial T}(1 - x) = \frac{\partial k_r}{\partial T}(C_{B_0} + C_{A_0}x)$$

From the Arrhenius equation, $\partial k_i/\partial T = k_i(E_i/RT^2)$, so

$$\frac{C_{A_0}(1 - x)}{C_{B_0} + C_{A_0}x} = \left(\frac{k_r^\circ e^{-E_r/RT}}{k_f^\circ e^{-E_f/RT}}\right)\left(\frac{E_r}{E_f}\right) \qquad (1\text{-}144)$$

For a given conversion, x, equation (1-144) defines a corresponding temperature maximizing the rate of reaction, which can then be used for calculation of the time of reaction from equation (1-142). In this case the temperature schedule would call for very high temperatures at the start of the reaction, maximizing the conversion of A while its concentration is high and that of product low, with a subsequent rapid decrease in temperature to prevent the reverse reaction from occurring, as shown in Figure 1.11.

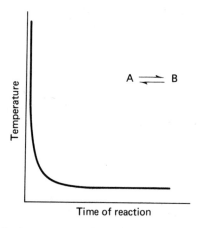

Figure 1.11 Optimal temperature schedule for reversible reactions, $E_r > E_f$

We shall see later that the time of reaction here corresponds to the size of a chemical reactor; thus what we have accomplished is the minimization of size and cost of the reactor system required to accomplish a specified conversion. Of course, absolute optimal temperature schedules may not always be practically accomplished, but they can serve at least as indicators of desirable or undesirable operational or design policies.

c. Interpretation methods—additional techniques

Most often for more complicated reactions than those discussed above, such as illustrated in Table 1.1b, direct methods of fitting conversion-time data become both tedious and difficult because of the awkward form of conversion equations and the number of rate constants involved in them. In such cases (and for the simpler ones also) the experimentalist can devise a number of special experiments which will simplify the subsequent task of interpretation. The most important of these are:

1. *Pseudo-order reactions:* When more than one component is involved in a reaction step (i.e., forward or reverse), one can separately identify order with respect to the individual components by conducting experiments in which the initial concentration of one is set far in excess of the second. Consider the reaction $aA + bB \rightarrow$ products. Then

$$\frac{dC_A}{dt} = -kC_A^p C_B^q \qquad \frac{dC_B}{dt} = -\nu_B k C_A^p C_B^q \qquad (1\text{-}152)$$

where ν_B is the ratio of stoichiometric coefficients, and the rate constant is based on the reaction rate of A. Now, if $C_{A_0} \gg C_{B_0}$, then the change in C_B is large compared to that in C_A, at least for small conversions, and the rate law effectively becomes

$$\frac{dC_B}{dt} = k'C_B^q \qquad (1\text{-}153)$$

The values of k' and q can be obtained from conversion-time data by methods already discussed. A second series of experiments is now run in which $C_{B_0} \gg C_{A_0}$, for which:

$$\frac{dC_A}{dt} = k''C_A^p \qquad (1\text{-}154)$$

The values of k'' and p are evaluated, and thence the value of the true rate constant k from knowledge of p, q, ν_B and the initial concentrations used in the two series of experiments. The kinetics given by equations (1-153) and (1-154) are said to be pseudo qth- or pth-order in B and A, respectively. An example of this procedure is given in Figure 1.15 for the case of a simple second-order reaction. In this case $p = q = 1$ and the full process just described is not necessary; however, the two experiments provide replicate measurements of the rate constant k. Of course, a deviation from the straight-line plot in either experiment would be indicative of p or $q \neq 1$. It is also wise to vary the levels of C_{A_0} (for $C_{A_0} \gg C_{B_0}$) and C_{B_0} (for $C_{B_0} \gg C_{A_0}$) to check the consistency of the interpretation of reaction order obtained from a single pair of experiments.

2. *Initial rates:* One of the requirements of the pseudo-order reaction was that the conversion be small, at least in the sense that the variation

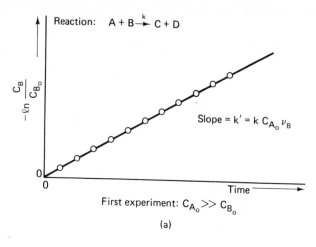

Reaction: $A + B \xrightarrow{k} C + D$

Slope = $k' = k\, C_{A_0}^{\nu_B}$

Time

First experiment: $C_{A_0} \gg C_{B_0}$

(a)

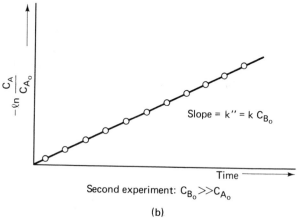

Slope = $k'' = k\, C_{B_0}$

Time

Second experiment: $C_{B_0} \gg C_{A_0}$

(b)

Figure 1.15 Example of pseudo-order experiment for a second-order reaction

in one term be negligible compared to the other. Rates of reaction at low conversions are called *initial rates* and correspond to the situation depicted on Figure 1.16. Under these conditions great simplifications can be made in kinetic expressions for complex reactions, particularly for reversible reactions. Consider the reaction $aA + bB \rightleftharpoons cC + dD$, for which a general rate expression would be

$$\frac{dC_A}{dt} = -k_f C_A^p C_B^q + k_r C_C^u C_D^v \qquad (1\text{-}155)$$

Now, if one conducts experiments under conditions such that C_{C_0} and C_{D_0} are zero, and the conversion of A is small, then

$$\frac{dC_A}{dt} = -k_f C_A^p C_B^q \qquad (1\text{-}152)$$

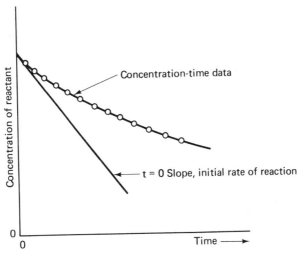

Figure 1.16 Initial rate of reaction as determined from slope of tangent to conversion-time curve at zero conversion

and we have the same problem illustrated in the section above. A similar technique may be applied to the reverse reaction, of course. With no A or B present initially,

$$\frac{dC_C}{dt} = -v_C k_r C_C^u C_D^w \tag{1-156}$$

where $v_C = c/a$ and the rate constants k_f and k_r are based on the reaction of A.

3. *Definition of the maxima of intermediates:* The effectiveness of this method has already been demonstrated for the Type III system and arises from the fact that the net reaction rate of intermediate is zero at the maximum, allowing one to work directly with the rate equation rather than the integral conversion-time relationship. See also problem 21.

By various combinations of these three methods, the kinetic constants of even very complex reactions can be obtained. At the least, one can expect to get preliminary values which can then be improved in obtaining a fit to the conversion-time data.

There are also other techniques for treating experimental data, and other specific experiments possible, than the simple methods discussed here. Measurements of reaction half-life, as defined in equations (1-45), are directly related to rate constants and also provide a simple means for determination of reaction order. From the expression for irreversible reaction of order $n \neq 1$:

$$t_{1/2} = \frac{2^{n-1} - 1}{k(n - 1)C_{A_0}^{n-1}} \tag{1-45a}$$

For $n = 1$,

$$t_{1/2} = \frac{\ell n\, 2}{k} \tag{1-45b}$$

For all values of n, $t_{1/2}$ can be written in general as

$$t_{1/2} = \frac{f(n, k)}{C_{A_0}^{n-1}} \tag{1-157}$$

or

$$\ell n\, t_{1/2} = \ell n\, f(n, k) - (n - 1)\, \ell n\, C_{A_0} \tag{1-158}$$

where $f(n, k)$ is some functional relationship pertaining to given kinetics and is a constant for constant temperature. Experimentally, we conduct two runs at the same temperature but different initial concentrations, $(C_{A_0})_1$ and $(C_{A_0})_2$, measuring the half-life in each. For each of these a relationship of the form of equation (1-158) exists; if the two are subtracted, $f(n, k)$ disappears since it is constant for the given temperature, and we may solve directly for n:

$$n = 1 + \frac{\ell n\, (t_{1/2})_2 - \ell n\, (t_{1/2})_1}{\ell n\, (C_{A_0})_1 - \ell n\, (C_{A_0})_2} \tag{1-159}$$

Graphically, a plot of $\ell n\, t_{1/2}$ versus $\ell n\, C_{A_0}$ should be a straight line according to equation (1-158), and the order determined directly from the slope, as shown in Figure 1.17. Similar methods of interpretation may be devised employing other fractional life periods with differing initial concentrations, or successive time intervals in a single run. Remember that half-life refers to the time required for half the initial reactant to be consumed, not to half the total time of reaction.

If direct information on rates of reaction is available, many of the tasks of interpretation are simplified, since one may work directly with the rate equations. The information required here is not concentration versus time, but rate of reaction versus concentration. As will be seen later, some types of chemical reactors give this information directly, but the constant-volume, batch systems discussed here do not. In this case it is necessary to determine rates from conversion-time data by graphical or numerical methods, as indicated for the case of initial rates in Figure 1.16. In Figure 1.18 is shown a curve representing the concentration of a reactant A as a function of time, and we identify the two points C_{A_1} and C_{A_2} for the concentration at times t_1 and t_2. The mean value for the rate of reaction we can approximate algebraically by

$$\frac{d\bar{C}_A}{dt} \approx \frac{C_{A_2} - C_{A_1}}{t_2 - t_1} \tag{1-160}$$

or can determine graphically from the slope of the tangent to the curve at the midpoint of the interval. The two methods do not generally give the same result, as Figure 1.18 illustrates. The success of such approximation obviously depends critically upon the precision of the data and the size of the interval,

his half-lives and he says that two half-lives are just about right. You may assume that he starts with the same sucrose concentration as in part (a) and that the rate constant is proportional to the weight ratio of yeast to sucrose.

17. The reaction

$$CH_3-C\overset{O}{\underset{O-CH_3}{\big\backslash}} \ + H_2O \longrightarrow CH_3-C\overset{O}{\underset{OH}{\big\backslash}} \ + CH_3OH$$

has been shown to be autocatalytic in acetic acid. The following data have been reported:

$$C_0(\text{acetate}) = 0.5 \text{ g mole/liter}$$

$$C_0(\text{acetic acid}) = 0.05 \text{ g mole/liter}$$

$$T = 40°C$$

After 1 hr the acetate concentration is 0.2 g mole/liter. At what time does the rate reach a maximum?

18. A group of workers investigating the reaction:

$$A \rightleftharpoons 2B \qquad (k_1 \text{ forward, } k_2 \text{ reverse})$$

under isothermal, constant-volume conditions report the equilibrium conversion to B as the following:

$$(C_B)_{eq} = \frac{1}{2k_2}[-(2k_2 C_{A_0} + k_1) + \sqrt{(2k_2 C_{A_0} + k_1)^2 + 8k_1 k_2 C_{A_0}}]$$

where C_{A_0} is the initial concentration of A and no B was present initially. The statement is made that "order conforms to stoichiometry"; prove or disprove this statement.

Section 1.5

19. The following reaction is carried out under isothermal, constant-volume conditions:

$$A \xrightarrow{k_1} B \underset{k_4}{\overset{k_3}{\rightleftharpoons}} C$$

Suppose that the starting charge is $C_{A_0} + C_{B_0}$, with C_{B_0} having the specific value $(C_B)_{eq}$, the equilibrium concentration. Thus, B starts and ends at the same concentration. What is the necessary value of C_{B_0} in terms of the reaction velocity constants and C_{A_0}? What relationships among k_1, k_3, and k_4 are required for a maximum of B to exist during the course of the reaction? Order conforms to stoichiometry.

20. The Type III reaction system discussed in the text consisted of irreversible individual steps; however, a more realistic model for series reactions might be:

$$A \underset{k_2}{\overset{k_1}{\rightleftharpoons}} B \underset{k_4}{\overset{k_3}{\rightleftharpoons}} C$$

Derive an expression for the concentration of B as a function of time of reaction under isothermal, constant-volume conditions. (*Hint:* Use the second derivative of C_A with respect to time of reaction.) A general method for solution of series first-order reversible reactions is given by A. Rakowski, *Z. Physik. Chem.*, **57**, 321 (1907).

21. Successive substitution reactions are very important in some organic syntheses. A good example is the chlorination of methane:

$$CH_4 + Cl_2 \longrightarrow CH_3Cl + HCl \qquad (k_1)$$

$$CH_3Cl + Cl_2 \longrightarrow CH_2Cl_2 + HCl \qquad (k_2)$$

$$CH_2Cl_2 + Cl_2 \longrightarrow CHCl_3 + HCl \qquad (k_3)$$

$$CHCl_3 + Cl_2 \longrightarrow CCl_4 + HCl \qquad (k_4)$$

Derive the expressions that relate the concentrations of the mono- and dichloro intermediates to the amount of methane reacted in an isothermal, constant-volume system. The initial concentrations of methane and chlorine are C_{M_0} and C_{C_0}, respectively, and no products are present initially. (*Hint:* Eliminate the time variable from the rate equations.) At what conversion do maxima in these intermediates occur if $k_1 = k_2 = k_3 = k_4$?

22. A kinetic model for the catalytic cracking of gas oil cycle stocks has been proposed as follows [V. W. Weekman, Jr., *Ind. Eng. Chem. Proc. Design Devel.*, **7**, 90 (1968); **8**, 388 (1969); V. W. Weekman, Jr. and D. M. Nace, *Amer. Inst. Chem. Eng. J.*, **16**, 397 (1970)]:

$$\text{gas oil} \longrightarrow \text{gasoline} \longrightarrow \text{dry gas} + \text{coke}$$

which can be represented roughly by the following sequence:

$$2A \xrightarrow{k_1} B \xrightarrow{k_2} C$$

with order corresponding to stoichiometry, and k_1 defined per mole of A reacted. Derive an expression for the yield and selectivity of gasoline formation according to this model. Assume isothermal reaction with constant volume.

23. Consider the successive reactions:

$$A + B \xrightarrow{k_1} C + D$$

$$C + B \xrightarrow{k_2} E + D$$

Derive the expression by which the ratio (k_2/k_1) can be determined from experimental conversion data (i.e., products C and E versus A). The initial concentrations of C and E are zero. For the special case $C_{A_0} = 400$, $C_{B_0} = 300$, $(k_2/k_1) = 0.5$, what is the ratio of E to C at the completion of the reaction?

24. (a) It is stated in the text that differential selectivity for the Type II system is defined in terms of consumption of reactant in order to work with positive numbers. Yet, an analogously defined differential selectivity for a Type III reaction gives negative values for certain ranges of intrinsic selectivity and reaction mixture composition, shown in Figure 1.9. What is the significance of these negative values?

(b) Derive equation (1-76) for the yield in a Type II system.

$$k = k° e^{-E/RT}$$

with the approximation

$$k = a° \left(\frac{T}{T_0}\right)^n$$

for this problem.

33. Direct combination of nitrogen and oxygen is carried out at 4000°F until equilibrium is attained; then the gases are cooled quickly to retard the reverse reaction. The reaction is second-order in both directions, with the rate equation

$$r = \frac{d(P_{NO})}{dt} = k_1 \left(P_{N_2} P_{O_2} - \frac{P_{NO}^2}{K}\right)$$

The rate constant is given as

$$\log_{10}\left(\frac{k_1}{3}\right) = \frac{T - 3800}{300}$$

$$T = °F$$

$$k_1 = (\text{atm-sec})^{-1}$$

Determine the mole fraction of NO present versus time and temperature for a cooling rate of 30,000°F/sec, starting at 4040°F and operating at atmospheric pressure. The following data are available for the equilibrium constant K:

T (°F)	1700	2060	2420	2780
$K = \dfrac{P_{NO}}{[P_{O_2} P_{N_2}]^{1/2}}$	5.26×10^{-4}	1.92×10^{-3}	5.08×10^{-2}	1.08×10^{-2}
T (°F)	3140	3500	3860	4040
$K = \dfrac{P_{NO}}{[P_{O_2} P_{N_2}]^{1/2}}$	1.98×10^{-2}	3.25×10^{-2}	4.91×10^{-2}	5.89×10^{-2}

34. While it is relatively simple to define the trends in point or differential selectivity with temperature in the examples discussed in the text, the trends of yield and overall selectivity are more complicated, since their expressions are not conveniently written in terms of rate-constant ratios. In batch operation, however, these are more pertinent measures of efficiency than point selectivity. Demonstrate the dependence of the yield of B, Y_B(III) [equation (1-91)] for systems in which $E_1 > E_2$ and $E_1 < E_2$ in the special case where $k_1° = k_2°$.

35. Borchardt and Daniels [H. J. Borchardt and F. Daniels, *J. Amer. Chem. Soc.*, **79**, 41 (1957)] report a simple means for determining the kinetics of irreversible reactions using differential thermal analysis. Although specific designs of DTA equipment vary, generally the reactive substance (solution) to be studied is placed in a sample cell and an inert solution in a reference cell. The inert solution is selected to have approximately the same thermal conductivity and heat capacity as the reactant solution, and the two cells are so designed and agitated that equal heat-transfer coefficients will be effective when the cells are immersed in a temperature bath. Within the bath, the temperature is normally changed such that the temperature of the reference cell, T_r, increases linearly with time. In the absence of reaction the sample temperature, T, will also increase linearly and equal T_r. When an exothermic reaction occurs in the sample cell, the tempera-

ture T will be greater than T_r, and the difference, $\Delta T = T - T_r$, can be recorded, as well as the actual reference temperature, T_r. The outcome of the experiment is a record of ΔT versus time, as shown in Figure 1.19.

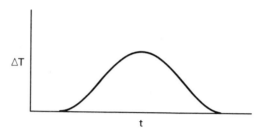

Figure 1.19 Typical result of a DTA experiment for an irreversible exothermic reaction

(a) From a heat balance on the system, show how the heat of reaction can be determined from the DTA curve. Use the following nomenclature:

C_p = specific heat of reaction or reference mixture
U = heat-transfer coefficient for total cell area
$$\alpha = \frac{C_p}{U}$$
k = rate constant in: $\dfrac{d(N/V)}{dt} = k\left(\dfrac{N}{V}\right)^n$
N = moles reactant (N_0 at $t = 0$)
V = volume of reactant solution
n = order of reaction
t = time

(b) Using the result of the heat-of-reaction determination, eliminate this quantity and N (since this is not measured in DTA) from your heat balance. Derive an expression, making use of properties observable or calculable from the DTA graph, which can be employed to determine the rate constant (expressed as a frequency factor and activation energy) and the order of reaction from the results of a series of DTA experiments. Show how you would do this (the final form of equation will require assumption of one of the three parameters and subsequent checking).

36. The following data pertain to the first-order reversible reaction $A \rightleftharpoons B$:

$$k_f = 4 \times 10^{13} \exp\left(\frac{-25{,}000}{RT}\right) \text{ days}^{-1}$$

$$k_r = 6.67 \times 10^{14} \exp\left(\frac{-30{,}000}{RT}\right) \text{ days}^{-1}$$

$$T = {}^\circ K; \quad R = \text{cal/gmole-}{}^\circ K$$

Determine the equilibrium conversion at $510{}^\circ K$ for an initial composition of 1 unit of A and 0.1 unit of B. A conversion of 99.1% of this equilibrium value is sought. What is the optimum isothermal temperature (minimum reaction time) for this conversion.

37. Calculate the optimal temperature program, including temperature versus conversion and the absolute minimum time, for the reaction of problem 36.

38. Consider the reactions

$$NH_3 + CH_3OH \xrightarrow{k_1} CH_3NH_2 + H_2O$$

$$CH_3NH_2 + CH_3OH \xrightarrow{k_2} (CH_3)_2NH + H_2O$$

$$(CH_3)_2NH + CH_3OH \xrightarrow{k_3} (CH_3)_3N + H_2O$$

All are irreversible. Some laboratory data obtained under isothermal conditions are as follows:

$\dfrac{a - x}{a}$	$\dfrac{y_1}{a}$	$\dfrac{y_2}{a}$
0.9	0.10	0.02
0.8	0.18	0.03
0.7	0.26	0.04
0.6	0.34	0.06
0.5	0.39	0.11
0.4	0.43	0.21
0.3	0.45	0.22
0.2	0.43	0.30
0.1	0.35	0.39
0.05	0.26	0.45
0.01	0.11	0.40

where a is the initial concentration of NH_3, y_1 the concentration of CH_3NH_2, y_2 the concentration of $(CH_3)_2NH$, and $(a - x)/a$ the fraction NH_3 unconverted. From these data determine the relative values (k_2/k_1) and (k_3/k_1). How would you go about determining absolute values for k_1, k_2, and k_3?

39. Here are some data on the chlorination of toluene in 99.87% acetic acid at 25°C, as reported by Brown and Stock [H. C. Brown and L. M. Stock, *J. Amer. Chem. Soc.*, **79**, 5175 (1957)]:

Molar Concentrations

Time (sec)	Toluene	Chlorine
0	0.1908	0.0313
2,790	0.1833	0.0238
7,690	0.1745	0.0150
9,690	0.1719	0.0123
14,000	0.1682	0.0086
19,100	0.1650	0.0055

Determine a reasonable kinetic sequence to explain these data and the value(s) of appropriate constant(s). Reaction was done in the dark to prevent chlorination of the methyl group.

40. The reaction

$$CH_3COCH_3 + HCN \longrightarrow CH_3-\overset{\overset{\displaystyle OH}{\displaystyle |}}{\underset{\underset{\displaystyle CN}{\displaystyle |}}{C}}-CH_3$$

was carried out in aqueous solution by Svirbely and Roth [W. J. Svirbely and J. F. Roth, *J. Amer. Chem. Soc.*, **75**, 3106 (1953)]. Typical data were as follows:

$$\text{initial HCN concentration} = 0.0758 \ N$$

$$\text{initial acetone concentration} = 0.1164 \ N$$

Time (min)	Concentration of HCN
4.37	0.0748 N
73.23	0.0710 N
172.5	0.0655 N
265.4	0.0610 N
346.7	0.0584 N
434.7	0.0557 N

Determine a reasonable interpretation for the kinetics of this reaction with values of the appropriate rate constants.

41. The series reaction system:

$$2A \xrightarrow{k_1} B \xrightarrow{k_2} C$$

has been studied in a constant-volume, batch reactor under isothermal conditions, with the following results:

	Concentrations (mole/volume)		
Time (hr)	A	B	C
0	1.0	0	0
0.03	0.76	0.098	0.02
0.06	0.63	0.122	0.06
0.1	0.51	0.140	0.10
0.15	0.39	0.120	0.17
0.2	0.33	0.10	0.24
0.3	0.25	0.05	0.32

Assuming that the order conforms to the stoichiometry, what are the values of k_1 and k_2?

42. The following data were obtained in an experiment to determine the rate constant of a half-order irreversible reaction (no volume change).

$$\text{initial concentration} = 0.78 \ M$$

$$\text{concentration at } t = 15 \text{ min} = 0.43 \ M$$

Boltzmann's constant. Since the individual components of velocity are independent of each other, we may write from equation (2-5) for distribution of an individual component $P(v_i)\, dv_i$:

$$P(v_i)\, dv_i = \frac{1}{\alpha \pi^{1/2}} e^{-v_i^2/\alpha^2}\, dv_i \qquad (2\text{-}5a)$$

For our purposes it is most convenient to work with the velocity distribution without regard to individual component directions; this is the speed, c, as shown in Figure 2.3, and is defined by

$$c^2 = v_x^2 + v_y^2 + v_z^2 \qquad (2\text{-}7)$$

To use this quantity, spherical coordinates are required, as shown in Figure 2.4. Using the coordinate transformations

$$v_x = c \cos \phi \sin \theta$$
$$v_y = c \sin \phi \sin \theta$$
$$v_z = c \cos \theta$$

Equation (2-5) becomes

$$P(c, \theta, \phi)\, dc\, d\theta\, d\phi = \frac{c^2 \sin \theta}{\alpha^3 \pi^{3/2}} e^{-c^2/\alpha^2}\, dc\, d\theta\, d\phi \qquad (2\text{-}8)$$

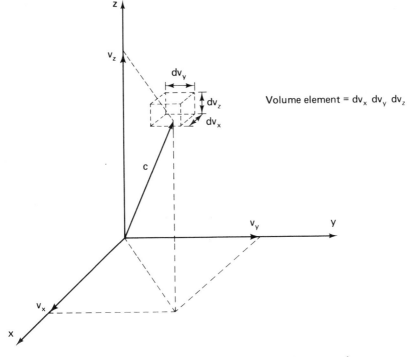

Volume element = $dv_x\, dv_y\, dv_z$

Figure 2.3 Speed and individual velocity components in rectangular coordinates

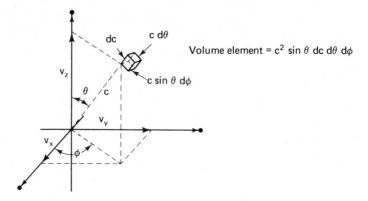

Figure 2.4 Speed in terms of spherical coordinates (after S. W. Benson, *The Foundations of Chemical Kinetics*, © 1960; with permission of McGraw-Hill Book Company, New York)

where $P(c, \theta, \phi)\, dc\, d\theta\, d\phi$ represents the fraction of molecules with velocity vectors in the range c to $c + dc$, θ to $\theta + d\theta$, and ϕ to $\phi + d\phi$. For an isotropic medium, we obtain the distribution of molecular speed by integrating equation (2-8) with respect to θ (0 to π) and ϕ (0 to 2π). The result is

$$P(c)\, dc = \frac{4c^2}{\alpha^3 \pi^{1/2}} e^{-c^2/\alpha^2}\, dc \tag{2-9}$$

The nature of these distribution functions, for an individual Cartesian velocity component and for molecular speed, is shown in Figure 2.5. The velocity component distribution is symmetric about the origin since the range of a velocity component is $-\infty$ to $+\infty$, while speed is a positive quantity.

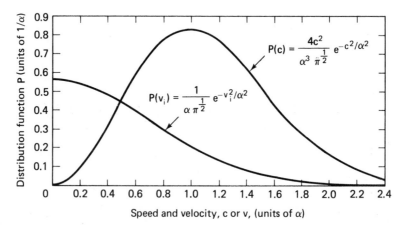

Figure 2.5 Cartesian velocity component and speed distribution functions (after S. W. Benson, *The Foundations of Chemical Kinetics*, © 1960; with permission of McGraw-Hill Book Company, New York)

that, by considering the rate of effusion of molecules in a vessel through a small orifice such that the velocity distribution of molecules in the vessel is not affected, one may derive

$$\bar{Z}_{cT} \text{ (surface)} = \tfrac{1}{4} n_A \left(\frac{8kT}{\pi m_A}\right)^{1/2} \qquad (2\text{-}23)$$

TABLE 2.2b
SPEEDS AND COLLISION NUMBERS FOR SOME TYPICAL GASES[a]

Gas	Molecular Weight	Diameter (Å) (from viscosity data)	$\bar{c}$ at 25°C (km/sec)	Mean Free Path (Å) (STP)	Collision Number $\times 10^{-28}$ (number/cm³-sec)
H_2	2.016	2.74	1.772	1180	20.4
He	4.002	2.18	1.257	1765	9.13
N_2	28.02	3.75	0.475	596	10.22
C_2H_6	30.05	5.30	0.448	298	19.7
O_2	32.00	3.61	0.434	644	8.87
Ar	39.94	3.64	0.3975	633	8.07
CO_2	44.00	4.59	0.379	397	12.22
Kr	82.9	4.16	0.276	485	7.32

[a]Relative speed for like molecules: $c_r = \sqrt{2}\, \bar{c}$

Average speed, $\bar{c} = \left(\frac{8kT}{\pi \mu}\right)^{1/2}$

Ideal gas number density (STP) = 2.687×10^{19} molecules/cm³
Electronic mass = 9.108×10^{-28} g
Avogadro's number = 6.025×10^{23} molecules/mole
Planck's constant (h) = 6.625×10^{-27} erg-sec/molecule
Gas constant (R) = 8.317×10^{7} ergs/mole-°K
Boltzmann constant ($k = R/N$) = 1.38×10^{-16} erg/molecule-°K
Source: After S. W. Benson, *The Foundations of Chemical Kinetics*, Table VII. 2, p. 155; © 1960, with permission of McGraw-Hill Book Company, New York.

2.2   Collision Theory of Reaction Rates

If we assume that every collision is effective in reaction, then $\bar{Z}_{cT}(A, B)$ from equation (2-21) gives us the rate of the reaction $A + B \rightarrow$ products directly. Previously, we have written for the irreversible, second-order reaction between A and B, for constant-volume conditions,

$$r_A = \frac{dC_A}{dt} = -k' C_A C_B \qquad (1\text{-}39)$$

For concentrations expressed as number densities (molecules/cm³), $C_A = n_A$, and so on, and the rate constant k' in molecular units, then, from equation (2-21), is

$$k' = \pi \sigma_{AB}^2 \left(\frac{8kT}{\pi \mu_{AB}}\right)^{1/2}$$

The absence of an activation-energy term in the expression for the rate constant is an immediate indication that the picture of each collision resulting in reaction is an incorrect one. Physical intuition tells us that it is more reasonable to expect reaction in a collision between two molecules with high energy (i.e., speed, since the hard-sphere molecule can possess only kinetic energy) than between two with low energy. In real reactions, this energy is required to overcome the repulsive forces (positive potential energies in Figure 2.1b) to allow the bonding distances of the product molecule (perhaps much smaller than the dimensions of the individual reactants) to be established. According to this picture, then, we are not interested in the total collision number but in a collision number that pertains to interactions in which energies greater than some threshold or minimum value are involved. To do this we introduce in the next section the concept of scattering cross sections.

a. Collision numbers in terms of relative kinetic energy; bimolecular reactions[1]

An alternative means to derive the result of equation (2-12) for collisions between a single A molecule and a matrix of B molecules is to consider a physical picture in which the A molecule is fixed in position and the B molecules approach in a beam of number density n_B, all with the same speed, c_B, and direction. The fixed molecule A presents an effective target area of σ_{AB}^2, and the moving molecules that are deflected per unit time have their centers within a circular cylinder of length c_B and area σ_{AB}^2. In this picture, σ_{AB}^2 is often referred to as the *sphere of influence* of the two molecules, and the moving B molecules are considered to be *scattered* by the fixed A. Note that the symbol σ_{AB} has a different meaning than that in Section 2.1.

The number of beam molecules, N, scattered in unit time by a fixed molecule is given by

$$N = \sigma I \tag{2-24}$$

where I is the intensity of the beam and σ the *cross section* of the sphere of influence. In the simple example of the previous paragraph:

$$I = n_B c_B$$

$$\sigma = \pi \sigma_{AB}^2$$

In the more general case, we will write equation (2-24) as

$$\sigma = \frac{N}{I} \tag{2-24a}$$

[1] The material of this section follows in general outline the treatment given by R. D. Present, *Kinetic Theory of Gases*, McGraw-Hill, New York, 1958, to which the reader is referred for a more complete discussion.

and use this to define the scattering cross section, σ, as the number of beam molecules scattered per time by one fixed molecule per unit intensity of the beam. The dimensions of σ will thus be "area."

As an example of the calculation of a cross section which will be useful to our collision-reaction theory, let us consider the collision between two spheres which possess weakly attractive forces with shorter range repulsion. The trajectory of a particle in the incident beam which is scattered is shown in Figure 2.7. In this case, the hard-sphere diameter cannot be used to define the scattering cross section because of the existence of attractive/repulsive forces; instead, we replace this with the quantity b, termed the *impact parameter*, representing the separation between the centers of the two particles. The other quantities shown on Figure 2.7 are χ, the angle of deflection; r and ϕ, the polar coordinates specifying the position of the particle with center of force at the origin; and r_m and ϕ_m, polar coordinates at the point of closest approach. The velocity of approach of the moving particle is v_0 at a point where there are no attractive/repulsive forces, and the linear momentum of approach is mv_0. Now angular momentum is conserved, so that

$$mv_0b = mr^2\dot{\phi} = p_\phi = \text{constant}$$

where the impact parameter b is taken as the "lever arm" in the conservation equation. In addition, total energy is conserved in the collision:

$$\frac{m(dr/dt)^2}{2} + \frac{p_\phi^2}{2mr^2} + u(r) = \frac{mv_0^2}{2} = \text{constant} \qquad (2\text{-}25)$$

where $u(r)$ denotes the potential-energy function describing the attractive–repulsive forces as a function of r.

We are interested first in determining r_m, the point of closest approach.

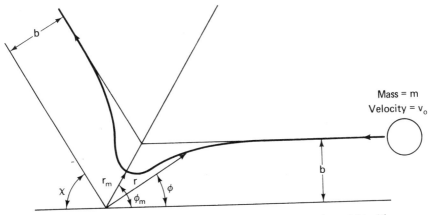

Figure 2.7 Trajectory of a particle in a centrally attractive force field with short-range repulsion (after R. D. Present, *Kinetic Theory of Gases*, © 1958; with permission of McGraw-Hill Book Company, New York)

This we can obtain by first noting that

$$\left(\frac{dr}{dt}\right) = 0 \qquad \text{at } r = r_m$$

so at this point equation (2-25) becomes

$$\frac{p_\phi^2}{2mr_m^2} + u(r_m) = \frac{mv_0^2}{2} \qquad (2\text{-}25a)$$

Solving for p_ϕ^2:

$$p_\phi^2 = m^2 r_m^2 v_0^2 - 2mr_m^2 u(r_m)$$

but $p_\phi^2 = m^2 v_0^2 b^2$, so

$$\frac{b^2}{r_m^2} = 1 - \frac{u(r_m)}{\frac{1}{2}mv_0^2} \qquad (2\text{-}26)$$

Equation (2-26) allows us to define the distance of closest approach in terms of the impact parameter and the potential function.

Just as we were able to convert equation (2-12) into the proper expression for a collision number on substituting c_r for c_A, we may write equation (2-26) for collision between moving molecules if we use relative velocity, v_r, in place of v_0. Further, if dissimilar molecules A and B are colliding, then μ_{AB} replaces m, and the distance of closest approach must be σ_{AB}. Then

$$b_{max} = \sigma_{AB}\left[1 - \frac{u(\sigma_{AB})}{E_r}\right]^{1/2}$$

where $E_r = \frac{1}{2}mv_r^2$. The physical interpretation of b_{max} is that only those molecules with centers within a circle of b_{max} will undergo collision (and reaction) for a given value of E_r. Hence the scattering cross section for this process is

$$\sigma = \sigma(E_r) = \pi b_{max}^2 = \pi \sigma_{AB}^2\left[1 - \frac{u(\sigma_{AB})}{E_r}\right] \qquad (2\text{-}27)$$

A physical picture of the potential function $u(\sigma_{AB})$ is given in Figure 2.8a for the limiting cases of attracting and repelling spheres. If η^* is the value of $u(\sigma_{AB})$, then

Attracting spheres:

$$\sigma(E_r) = \pi \sigma_{AB}^2\left(1 + \frac{\eta^*}{E_r}\right)$$

and

$$\sigma(E_r) \geq \pi \sigma_{AB}^2$$

Repelling spheres:

$$\sigma(E_r) = \pi \sigma_{AB}^2\left(1 - \frac{\eta^*}{E_r}\right) \qquad E_r > \eta^*$$

and

$$\sigma(E_r) \leq \pi \sigma_{AB}^2$$

or

$$\sigma(E_r) = 0 \qquad E_r < \eta^* \qquad \text{(no collision)}$$

in which bonds are formed rather than broken and in which the energetics of reaction are represented in terms of relative kinetic energy.

Let us look at a typical unimolecular reaction, for example the first step of the diethylether decomposition reaction illustrated in chapter 1:

$$C_2H_5OC_2H_5 \longrightarrow \cdot CH_3 + \cdot CH_2OC_2H_5$$

Here the reaction process is one of spontaneous transformation which requires sufficient energy to be present in the ether molecule to permit the rupture of a carbon/carbon bond. This energy is obviously internal to the molecule and cannot be represented as a translational-energy term. The pertinent question to ask is how a molecule acquires the required energy for the transformation to occur, and the answer lies in a consideration of the energy exchange from external (kinetic) to internal (rotational and vibrational) modes in polyatomic molecules.

We can obtain at least a qualitative picture of such exchange by a modification of our previous hard-sphere approach. First, recall that in equations (2-3) and (2-4) we obtained expressions for the postcollision velocities of two hard spheres in terms of the precollision values. The change in kinetic energies for this process is

$$\Delta E_A = -\Delta E_B = \tfrac{1}{2}m_A(v_A^2 - v_A'^2) \tag{2-36}$$

and the fraction of the original energy of sphere A that is exchanged is

$$\frac{\Delta E_A}{E_A} = \frac{\tfrac{1}{2}m_A(v_A^2 - v_A'^2)}{\tfrac{1}{2}m_A v_A^2} = 1 - \frac{v_A'^2}{v_A^2} \tag{2-37}$$

In terms of the precollision values of v_A and v_B:

$$\frac{\Delta E_A}{E_A} = \frac{4\theta}{(1+\theta)^2}\left(1 - \frac{v_B}{v_A}\right)\left(1 + \frac{1}{\theta}\frac{v_B}{v_A}\right) \tag{2-38}$$

where $\theta = m_A/m_B$. Now let us consider this collision to be that shown in Figure 2.9, in which one of the hard spheres is an atom in a diatomic molecule and only energy exchange along the line of centers is considered (i.e., a head-

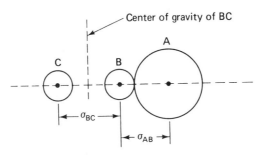

Figure 2.9 Collisions between an atom and a diatomic molecule (after S. W. Benson, *The Foundations of Chemical Kinetics*, © 1960; with permission of McGraw-Hill Book Company, New York)

on collision). The same relationship, equation (2-38), pertains to energy exchange between the atom and molecule, if we neglect any BC interactions. For the conservation of momentum in this collision, we can write

$$m_{BC}(v'_B - v_{BC}) = m_A(v_A - v'_A) \qquad (2\text{-}39)$$

and in terms of precollision volocities:

$$\frac{v'_{BC}}{v_A} = \frac{2m_B m_A}{m_{BC} m_{AB}}\left(1 - \frac{v_B}{v_A}\right) + \frac{v_{BC}}{v_A} \qquad (2\text{-}40)$$

The energy, ΔE_v, transferred to the internal mode (vibrational) of BC is the difference between $(\Delta E_A/E_A)$ and that appearing as a change in the translational energy of BC:

$$\frac{\Delta E_v}{E_A} = \frac{\Delta E_A}{E_A} - \frac{1}{2}\frac{m_{BC}(v'^2_{BC} - v^2_{BC})}{E_A} \qquad (2\text{-}41)$$

Substituting from equation (2-40) for v'_{BC}, we have

$$\frac{\Delta E_v}{\Delta E_A} = \frac{4m_C m_B m_A}{m^2_{AB} m_{BC}}\left(1 - \frac{v_B}{v_A}\right)\left(1 + \frac{v_B}{v_A}\frac{m_{ABC}m_B}{m_C m_A} - \frac{v_{BC}}{v_A}\frac{m_{AB}m_{BC}}{m_C m_A}\right) \qquad (2\text{-}42)$$

The value of $(\Delta E_v/\Delta E_c)$ may be either positive or negative here, depending on the relative signs of v_B, v_A, and v_{BC}, since the picture of hard-sphere collisions requires an instantaneous interaction. For real molecules, however, the collisions will take place over a finite period of time in which several vibrational cycles of the molecule BC occur (i.e., v_B and v_C changing directions with respect to each other), and we can time-average results such as that of equation (2-42).

The concept we are trying to illustrate from this simplified analysis is *collisional activation*, by which a polyatomic molecule through a series of favorable collisions is able to accumulate internal energy from kinetic-energy exchange. The concept of collisional activation was used by Lindemann [F. A. Lindemann, *Trans. Faraday Soc.*, **17**, 598 (1922)] as the basis for his original development of a theory for unimolecular reactions. Consider the unimolecular reaction A $\longrightarrow$ products, which we will write in two steps:

$$\begin{aligned} A + A &\rightleftharpoons A^* + A \qquad (k_1, k_{-1}) \\ A^* &\rightleftharpoons \text{products} \qquad (k_2) \end{aligned} \qquad \text{(XIII)}$$

The first step involves the activation of the A molecule by collision with another A molecule, or in general with any other molecule, M, in the system to give A*, and the reverse process of collisional deactivation of A*, while the second step is the decomposition of A*. If the deactivation rate is large compared to the decomposition we may apply the pssh for A*:

$$\frac{dC^*_A}{dt} = 0 = k_1 C^2_A - k_{-1} C^*_A C_A - k_2 C^*_A \qquad (2\text{-}43)$$

from which

$$C^*_A = \frac{k_1 C^2_A}{k_2 + k_{-1} C_A}$$

and

$$\text{rate} = k_2 C_A^* = \frac{k_1 k_2 C_A^2}{k_2 + k_{-1} C_A} \tag{2-44}$$

This expression has two limits. At sufficiently high concentrations, $k_{-1} C_A \gg k_2$ and

$$\text{rate} \approx \frac{k_1 k_2}{k_{-1}} C_A \doteq k_\infty C_A \tag{2-45}$$

and at very low concentrations, $k_{-1} C_A \ll k_2$, and

$$\text{rate} \approx k_1 C_A^2 \tag{2-46}$$

The rate of the unimolecular reaction is thus first order at high pressure and second order at low pressure, or if one interprets the results wholly on the basis of a first-order form,

$$r = k^1 C_A \tag{2-47}$$

then the value of k^1 decreases with pressure. This change has been observed experimentally for a number of reactions. Some typical data [H. O. Pritchard, R. G. Sowden, and A. F. Trotman-Dickenson, *Proc. Roy. Soc. (London)*, **217A**, 563 (1963)] for the isomerization of cyclopropane,

$$\underset{\underset{H_2C-CH_2}{\diagdown\diagup}}{\overset{\overset{H_2}{|}}{C}} \longrightarrow CH_3-\underset{H}{\overset{}{C}}\!=\!CH_2$$

are shown in Figure 2.10, where the ratio of k^1 to the limiting high-pressure rate constant, k_∞, is plotted versus cyclopropane pressure.

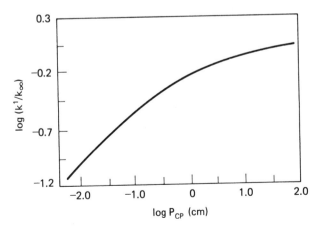

Figure 2.10 Pressure dependence of the first-order rate constant for cyclopropane isomerization (after H. O. Pritchard, R. G. Sowden, and A. F. Trotman-Dickenson, *Proc. Roy. Soc. (London)*, **A217**, 563 (1953); with permission of The Royal Society)

If we write k^1 in terms of the individual rate constants of equation (2-44):

$$k^1 = \frac{k_1 k_2 C_A}{k_{-1} C_A + k_2} \tag{2-48}$$

which can be rearranged to

$$k^1 = \frac{k_\infty}{1 + (k_\infty/k_1)(1/C_A)} \tag{2-49}$$

where according to a simple collision theory k_1 could be calculated from the collision number, $\bar{Z}_n(A, A)$. If this is done, however, one finds only qualitative agreement with the pressure variation of rate constant. As will be found in problem 10, the theoretical prediction tends to lie below the experiment over much of the pressure range. Now, if k^1 is decreasing from the high-pressure limit at too large values of C_A, it means that in equation (2-49) the denominator is too large, which can occur either because k_∞ is too large or k_1 is too small. Since k_∞ is the experimentally observed first-order constant at high pressure, however, the discrepancy can only be due to a value of k_1 which is too small. The reason for this lies in the fact that k_1 was computed from the collision-theory result for hard spheres, $\bar{Z}_n(A, A)$, and even though an activation energy is involved, the true activation must occur more rapidly if we include the influence of energy transfer from external to internal modes in the calculation for k_1. In the following section we discuss briefly several modifications of the Lindemann theory based on more realistic molecular models which lead ultimately to a workable theory for the kinetics of elementary steps.

c. Modifications of the Lindemann theory

The first modification, due to Hinshelwood [C. N. Hinshelwood, *Proc. Roy. Soc.* (*London*), **113A**, 230 (1927)], is to abandon the hard-sphere collision theory result in favor of a more detailed consideration of energy exchange in complex molecules. We have seen from equation (2-42) that even the simplest picture of a polyatomic molecule indicates that a significant fraction of translational energy can be changed to vibrational energy on collision. Let us now assume that the molecules being activated consist of n weakly coupled harmonic oscillators (no interaction between the oscillators), and that we wish to determine the distribution of such molecules with total energy between E and $E + dE$ with a fixed distribution $E_1, E_2, \ldots, E_n$ of energy between the oscillators, with $E = \sum_{i=1}^{n} E_i$. There exists a well-defined relationship between the number of oscillators, n, and the number of atoms in the molecule, N, since we can view n as representing the number of degrees of vibrational freedom in the molecule. For a molecule of N atoms, $3N$ coordinates are required to locate the positions of individual atoms, and since the atoms of a molecule move through space as a coherent group the motion of the molecule

The integral of equation (2-65) is the gamma function, defined as

$$\Gamma(n) = \int_0^\infty y^{n-1} e^{-y} \, dy \tag{2-66}$$

which, again, is a tabulated function. When n is an integer, however, $\Gamma(n) = (n-1)!$, and equation (2-65) gives a familiar looking result for k^1:

$$\bar{k}^1 = \bar{v} e^{-E^*/kT} \tag{2-67}$$

Since this corresponds to the high-pressure limit, $k^1 = k_\infty$.

At the low-pressure limit, $k_{-1} C_A \ll k_2(E)$ and equation (2-63) becomes

$$\bar{k}^1 = \frac{\bar{Z}'_{cT}(A, A^*) C_A}{(n-1)!} \int_{E^*}^\infty \left(\frac{E}{kT}\right)^{n-1} e^{-E/kT} \frac{dE}{kT} \tag{2-68}$$

which gives directly from the gamma function,

$$\bar{k}_1 = \bar{Z}'_{cT}(A, A^*) C_A = \pi \sigma_{AA}^2 \left(\frac{4kT}{\pi m_A}\right)^{1/2} C_A \tag{2-69}$$

The interpretation of this result is that deactivating collisions between A and A* are infrequent at low pressures; thus, the activated states have relatively long lifetimes and the apparent rate constant becomes equal to the rate of collisional activation of A.

It is seen that the theory also predicts the effect of pressure on the apparent activation energy, E_{app}. At high pressure, from equation (2-67),

$$E_{app} = kT^2 \frac{\partial \ln \bar{k}^1}{\partial T} = E^* \tag{2-70}$$

and at low pressure, from equation (2-69),

$$E_{app} = kT^2 \frac{\partial \ln \bar{k}^1}{\partial T} = \frac{1}{2} kT \tag{2-71}$$

We may draw several conclusions from the results of this example. First, comparison of equation (2-67) with the Arrhenius form suggests that the preexponential factor of the latter is a frequency related to the vibrational frequency of the oscillators in the molecule.[1] Such frequences as evaluated from molecular spectra are generally in the range 10^{12} to 10^{14} sec^{-1}; Arrhenius correlation of high-pressure rate constants yielding values much larger or smaller than this should be taken as indicating large structural changes occurring in the reaction. The reason for this and a more quantitative approach to such structural effects (basically, entropy factors) will be given in the next section. Second, the RRK rate constant in terms of average energy, equation (2-62), approaches the Arrhenius form as n becomes large. An illustration of this for $(E^*/kT) = 35$ is given in Figure 2.12. We also observe from Figure 2.12 that as the molecule becomes more complex (n increases) the ratio $(k_2(E_{av})/\bar{v})$ increases, meaning that the average lifetime of the

---

[1] More correctly, some characteristic frequency at which energy can be exchanged intramolecularly.

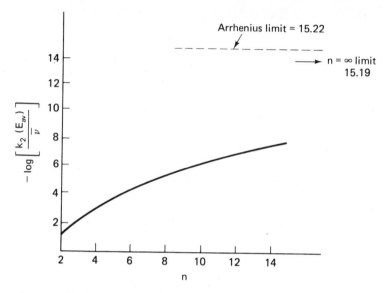

Figure 2.12 Comparison of $k_2(E_{avg})$ with the Arrhenius equation for $(E^*/kT) = 35$

molecule before decomposition increases with increasing molecular complexity. Finally, the different limiting forms for k^1 and the associated activation energies clearly indicate the changing nature of the controlling or dominant step in the reaction as the conditions of reaction change.

It is possible to visualize a large number of reactions in terms of collisional activation/unimolecular reaction steps similar to those involved in the basic Lindemann scheme. For example, a fully detailed model of the isomerization we have been discussing might be written

$$A \rightleftharpoons A^* \rightleftharpoons B^* \rightleftharpoons B \qquad (XIV)$$

where the first and last steps represent collisional activation and deactivation, and $A^* \rightleftharpoons B^*$ a true RRK step. At the next level of complexity, we might treat the bimolecular association reaction, $A + B \rightleftharpoons C$, by the following scheme:

$$
\begin{array}{ccc}
A + B & \rightleftharpoons & (AB) \\
& \searrow B^* + A & \\
\Updownarrow & \searrow & \Updownarrow \\
A^* + B & \rightleftharpoons & (AB)^* \rightleftharpoons C^* \rightleftharpoons C
\end{array}
\qquad (XV)
$$

where the unimolecular reaction step involves $(AB)^* \rightleftharpoons C^*$ and all others involve collisional activation or deactivation. In (XV), (AB) would represent an association complex of A and B without sufficient energy to react and (AB)* those complexes with sufficient energy for reaction, formed either from A* and B, B* and A, or collisional activation of (AB).

A number of cases such as this has been treated by Benson and Axworthy [S. W. Benson and A. E. Axworthy, Jr., *J. Chem. Phys.*, **21**, 428 (1953)], and

Unless one is dealing with excited states of molecules, a problem considerably beyond the scope of our present interests, electronic energy does not contribute to the partition function. The lowest-energy state of a molecule is ordinarily a singlet state; hence, $Q_e = 1$.

It is often convenient to represent these partition functions in terms of factors per degree of freedom. For an N-atom nonlinear molecule,

$$Q_t = f_t^3 \tag{2-85}$$

$$Q_r = f_r^3 \tag{2-86}$$

$$Q_v = f_v^{3N-6} \tag{2-87}$$

and

$$Q^\circ = f_t^3 f_r^3 f_v^{3N-6} \tag{2-88}$$

Such representation is useful when it is impossible or impractical to make detailed calculations of partition functions, but when order-of-magnitude estimates would be useful. In Table 2.3 is given a tabulation of representative

TABLE 2.3
ESTIMATED PARTITION FUNCTIONS
PER DEGREE OF FREEDOM[a]

Translation	f_t	10^8–10^9
Rotation	f_r	10^1–10^2
Vibration	f_v	10^0–10^1

$$^a f_t = \frac{(2\pi mkT)^{1/2}}{h}$$

$$f_r = \left(\frac{8\pi^2}{\sigma}\right)^{1/3} \frac{(8\pi^3 I_1 I_2 I_3)^{1/6}}{h}(kT)^{1/2}$$

$$f_v = (1 - e^{-h\nu_s/kT})^{-1}$$

Source: After A. A. Frost and R. G. Pearson, *Kinetics and Mechanism*, 2nd ed., © 1961; with permission of John Wiley & Sons, Inc., New York.

values for these factors as given by Frost and Pearson, computed from equations (2-80), (2-82), or (2-83) and (2-84) using typical values for molecular constants and a temperature range 300 to 500°K. It is seen that the translational contribution per degree of freedom is much larger than rotation or vibration; however, in complicated molecules it is possible for the *total* vibrational contribution, f_v^{3N-6}, to become large as $(3N-6)$ becomes a large number.

With these equations or estimates for the partition functions, we are now prepared to calculate the equilibrium constant, $K_c^\ddagger$, and thus in principle determine $C_\ddagger$ in terms of the observables C_{AB} and C_C and what we know or are prepared to postulate concerning the nature of the reactants and the transition complex.

c. The rate of reaction

We have stated the postulate of TST that the rate of reaction is the rate of passage of activated complexes through the transition state, that is, over the energy barrier shown in Figure 2.13. We can write this rate in terms of a frequency of passage over the barrier, v_t, such, that

$$r = v_t C_{\ddagger} \tag{2-89}$$

where r is the rate of reaction.

For our example reaction, if we substitute for $C_{\ddagger}$ from equation (2-77):

$$r = v_t \frac{Q_{\ddagger}^{\circ}}{Q_{AB}^{\circ} Q_C^{\circ}} e^{-E/RT} C_{AB} C_C \tag{2-90}$$

Now we must examine in detail the nature of the partition functions Q_{AB}°, Q_C°, and $Q_{\ddagger}^{\circ}$. The first two of these refer to the reactant molecules considered at equilibrium, and present no problem. However, $Q_{\ddagger}^{\circ}$ refers to the complex in the transition state, and it is reasonable to assume that the configuration of this state as required for reaction to occur may place some restriction on the number of degrees of freedom allowed, and this is indeed so. We have seen in Figure 2.13 that the relative values of R_{AB} and R_{BC} can serve as a measure of the extent of reaction (i.e., progress along the reaction coordinate, *not* conversion). If we view the passage through the transition state as a particular type of vibration, one in which C approaches B while A is departing, we may consider that one of the vibrational degrees of freedom is fixed by this movement through the transition state. This is of low frequency compared to ordinary molecular vibrations and we may write equation (2-84) for this particular vibration:

$$(f_v^{\ddagger})_i = (1 - e^{-hv_i/kT})^{-1} \tag{2-91}$$

which for low frequency is

$$(f_v^{\ddagger})_i = \frac{kT}{hv_i} \tag{2-92}$$

As the subscript in the equations above indicates, this frequency must also be identical with the frequency of passage through the transition state, so the rate equation becomes

$$r = \frac{(Q_{\ddagger}^{\circ})'}{Q_{AB}^{\circ} Q_C^{\circ}} \left(\frac{kT}{h} \right) e^{-E/kT} C_{AB} C_C \tag{2-93}$$

in which

$$(f_v^{\ddagger})_i (Q_{\ddagger}^{\circ})' = Q_{\ddagger}^{\circ} \tag{2-94}$$

In evaluation of the partition function of the transition complex, then, one must subtract one degree of vibrational freedom to account for motion along the reaction coordinate. For a nonlinear transition state,

$$(Q_{\ddagger}^{\circ})' = f_v^{3N-7} f_t^3 f_r^3 \tag{2-95}$$

TABLE 2.5g
IONIC REACTIONS IN AQUEOUS SOLUTION

Reaction	k° (liters/mole-sec)	$(\Delta S^\circ)^\ddagger$ (eu)
$Cr(H_2O)_6^{3+} + CNS^-$	1.3×10^{14}	0.7
$Co(NH_3)_5Br^{2+} + OH^-$	4.2×10^{17}	20.1
$ClO^- + ClO^-$	9.6×10^8	-19.6
$ClO^- + ClO_2^-$	8.5×10^8	-19.8
$CH_2BrCOO^- + S_2O_3^{2-}$	1.2×10^7	-28.3
$CH_2ClCOO^- + S_2O_3^{2-}$	2.3×10^9	-17.7
$Co(NH_3)_5Br^{2+} + Hg^{2+}$	1.3×10^8	-23.6
$S_2O_4^{2-} + S_2O_4^{2-}$	1.8×10^4	-41.2
$S_2O_3^{2-} + SO_3^{2-}$	2.3×10^6	-31.4

Source: After A. A. Frost and R. G. Pearson, *Kinetics and Mechanism,* 2nd ed., © 1961; with permission of John Wiley & Sons, Inc., New York.

TABLE 2.5h
VOLUMES AND ENTROPIES OF ACTIVATION

Reaction	$(\Delta V^\circ)^\ddagger$ (cm^3/mole)	$(\Delta S^\circ)^\ddagger$ (cal/°K-mole)
$CH_2BrCOOCH_3 + S_2O_3^{2-}$ $CH_2(S_2O_3)COOCH_3 + Br^-$	3.2	6
Sucrose $+ H_2O \xrightarrow{H^+}$ Glucose $+$ Fructose	2.5	8
$CH_2ClCOO^- + OH^-$ $CH_2OHCOO^- + Cl^-$	-6.1	-12
$CH_3CONH_2 + H_2O \xrightarrow{OH^-}$ $CH_3COOH + NH_3$	-14.2	-34

Source: After K. J. Laidler, *Chemical Kinetics,* 2nd ed., © 1965; with permission of McGraw-Hill Book Company, New York.

Laidler. It is important to appreciate the fact that the order of reaction given in the tables may not correspond to the true molecularity of the reaction, and thus in general we cannot claim to be quoting results for true elementary steps. However, order is a convenient means for classification, and in at least some cases (Tables 2.5c and d) a fair argument may be made that the results do pertain to elementary steps. The activation energy and preexponential factor values given in the table are from the Arrhenius correlation,

$$k = k^\circ e^{-E/RT} \qquad (1\text{-}25)$$

and where $(\Delta S^\circ)^\ddagger$ values depend on the choice of standard state (i.e., second-order reactions), a condition of 1 mole/cm^3 has been used.

The first-order reactions of Tables 2.5a and b are most probably not elementary steps, but many of them do probably involve some type of uni-molecular rearrangement as the first, and slow, step. The values of $(\Delta S^\circ)^\ddagger$

are evaluated by the procedure suggested in problem 19b, with positive or negative values corresponding to rates greater or less than the "norm", $(kT/h) = 10^{13}$, respectively. A number of the reactions in Table 2.5a have preexponential factors quite comparable to the norm, given the precision of most kinetic measurements,[1] and hence are in good agreement with order-of-magnitude theoretical results. On the other hand, note the very large negative values for $(\Delta S°)^{\ddagger}$ for first two isomerizations in Table 2.5b.

The radical recombination results of Table 2.5c are based on the data given for the methyl-radical recombination reaction. This reaction requires no activation energy and has a steric factor of unity, so the rate is given by the collision number in the system. Very little activation energy is required for any of these reactions and, again taking into account the precision of measurement, it would appear that the rates are all about the same order of magnitude. In Table 2.5d the steric factors reported are also determined from the collision-theory development of section 2.2; some of these will be recognized as the elementary steps of the chain reaction examples of chapter 1. The precision of these data is severely limited by the difficulties associated with determining free-radical concentrations, since they can only be determined indirectly (see Laidler, chapter 4, for a discussion of this).

The values for $(\Delta S°)^{\ddagger}$ in Table 2.5e are obtained from preexponential factor measurements, values for p determined from equation (2-33a) or (2-34a), and the estimation procedure suggested in problem 19a. In general, these are not elementary steps, although for many years it was believed that the hydrogen/iodine reaction was a true bimolecular reaction, since both collision theory and TST estimates of preexponential factor were in good agreement with experiment. This view, however, has changed [J. H. Sullivan, *J. Chem. Phys.*, **46**, 73 (1967)].

As mentioned earlier, the termolecular reactions, shown in Table 2.5f, are either NO reactions or atom recombinations involving a third-body collision. The activation energies in these reactions are all essentially zero and, indeed, one would not expect to observe termolecular reactions with significant activation energies. Since the collision numbers are very low compared with bimolecular values, the existence of a large activation energy, further decreasing the rate, would probably mean that some other mechanism of reaction would be favored.

In Tables 2.5g and h are reported some kinetics of reactions in which nonidealities are important; the former gives information on some typical ionic reactions in aqueous solution, and the latter on the effect of pressure on the rates of liquid-phase reactions. The large negative values for $(\Delta S°)^{\ddagger}$, corresponding to reaction between like charged ions in solution, is often the result of an effect known as *electrostriction*. The charge on the transition

---

[1] Frost and Pearson suggest, for example, that activation energies are normally not determined to better than kilocalorie accuracy and preexponential factors to better than within a factor of 4 or 5. "It is better to have a little than nothing" (Publilius Syrus).

complex is the algebraic sum of the charges on reactants; when these do not counteract each other, strong electrostatic forces are established in the solvent molecules surrounding the transition complex and the freedom of motion of that complex is severely restricted, hence a large decrease in entropy is associated with formation of the transition state. The activation volume data of Table 2.5h (see problem 21) are representative; the rate constant increases with pressure for negative $(\Delta V°)^{\ddagger}$, meaning that the transition state has a smaller volume than the reactants, and vice versa. It has been observed [M. W. Perrin, *Trans. Faraday Soc.*, **34**, 144 (1938)] that entropies and volumes of activation tend to correspond in sign, although solvent interactions may vary this relationship for the same reaction carried out in different solvents.

2.5   Estimation of Activation Energies

At several places in this discussion of reaction mechanisms we have made the point that while theory has much to say concerning preexponential factors, it is relatively uninformative with respect to activation energies. That this is so is not surprising; we have already commented upon what must be involved in determination of potential-energy surfaces such as that of Figure 2.14.

There are a number of semiempirical methods which have been proposed for estimation of activation energy, determined mostly in the light of correlation of experimental results. The most recent of these is provided by Benson (S. W. Benson, *Thermochemical Kinetics: Methods for the Estimation of Thermochemical Data and Rate Parameters*, 2nd ed., John Wiley, New York, 1976), who gives extensive compilations for gas-phase reactions, classified as to whether unimolecular (fission of one or two bonds, isomerization—various transition-state configurations), bimolecular (metathesis, displacement or exchange, association), or complex (chains).

Before entering into the very detailed approaches developed by Benson, one might first wish to explore the results provided by two earlier workers for several classes of reaction. Semenov (N. N. Semenov, *Some Problems in Chemical Kinetics and Reactivity*, vol. I, Princeton University Press, Princeton, N.J., 1958; translated by M. Boudart) has suggested the following for addition or abstraction reactions of atoms and small radicals:

$$E = 11.5 - 0.25(-\Delta H_R) \quad \text{(exothermic)}$$
$$E = 11.5 + 0.75(-\Delta H_R) \quad \text{(endothermic)}$$

(2-117)

where the activation energy E and the heat of reaction $(-\Delta H_R)$ are in kcal/mole for the particular values of the constants given. These correlations are particularly useful in estimation of energy requirements in screening of candidate reactions as elementary steps in chain reactions.

Earlier, Hirschfelder [J.O. Hirschfelder, *J. Chem. Phys.*, **9**, 645 (1941)] had provided some guides for estimation of activation energies in methathesis reactions of the form $A + BC \rightarrow AB + C$. When written in the exothermic direction, E is of the order of 5.5% of the dissociation energy of the bond broken; when written in the endothermic direction E is this amount plus the heat of reaction, $(-\Delta H_R)$. For displacement reactions of the type $AB + CD \rightarrow AC + BD$ it was proposed that E would be of the order of 28% of the sum of the bond energies of the bonds being broken (exothermic direction), or this amount plus the heat of reaction (endothermic direction).

A number of other correlations have been proposed, but these should suffice for this entering level discussion. They should be used with caution; Benson reports, for example, that neither of the two Hirschfelder rules is better than about ± 5 kcal/mole, and occasionally much larger discrepancies are encountered. Hence, we take these as qualitative norms to provide guidance rather than quantitative numbers to provide definition.

2.6   Linear Free-Energy Relationships:
Hammett and Taft Correlations

Some useful correlations for both rate and equilibrium constants have been obtained for homologous series of reactions involving related compounds in a number of instances. The best known of these, due to Hammett [L. P. Hammett, *Physical Organic Chemistry*, McGraw-Hill Book Company, New York, 1940; see also H. H. Jaffe, *Chem. Rev.*, **53**, 191 (1953)], deals with the reactions of various meta- and para-substituted benzene derivative compounds. The correlation proposes a linear log–log relationship between the rate or equilibrium constant of a given reaction and some reference reaction for the series, k', K, k'_0, and K_0, respectively. Hence,

$$\log k' = \log k'_0 + \sigma\rho \tag{2-118}$$

or

$$\log K = \log K_0 + \sigma\rho \tag{2-119}$$

In the sense of the original correlation, these pertained specifically to compounds with the same reactive group as a side chain with a series of substituents meta or para to it. The constants σ and ρ depend upon different factors: σ on the substituent group and ρ on conditions of reaction and solvent.

Such a correlation actually implies a linear relationship between the free energies of the reactions in the series. From the thermodynamic formulation of transition-state theory, we have

$$k' = \left(\frac{kT}{h}\right)e^{-\Delta G^{\ddagger}/RT}$$

so

$$\log k' = \log\left(\frac{kT}{h}\right) - \frac{\Delta G^{\ddagger}}{2.3RT} \qquad (2\text{-}120)$$

Equating (2-118) and (2-120) gives

$$\log k'_0 + \sigma\rho = \log\left(\frac{kT}{h}\right) - \frac{\Delta G^{\ddagger}}{2.3RT} \qquad (2\text{-}121)$$

or

$$\begin{aligned}\Delta G^{\ddagger}_1 &= (\Delta G^{\ddagger}_0)_1 - 2.3RT\rho_1\sigma_1 \quad \text{(reaction series 1)}\\[4pt] \Delta G^{\ddagger}_2 &= (\Delta G^{\ddagger}_0)_2 - 2.3RT\rho_2\sigma_2 \quad \text{(reaction series 2)} \end{aligned} \qquad (2\text{-}122)$$

The free-energy relationships for the two series can be combined to give

$$\Delta G^{\ddagger}_1 - \frac{\rho_1}{\rho_2}(\Delta G^{\ddagger}_2) = (\Delta G^{\ddagger}_0)_1 - \frac{\rho_1}{\rho_2}(\Delta G^{\ddagger}_0)_2 \qquad (2\text{-}123)$$

which is the linear relationship sought.

In the use of the Hammett correlation a value of unity is assigned to ρ for the ionization constant for benzoic acid in aqueous solution. From this

$$\sigma = \log\left[\frac{K_i\,(\text{substituted benzioc acid})}{K_i\,(\text{benzoic acid})}\right] \qquad (2\text{-}124)$$

Now, from σ defined by equation (2-124), values of ρ can be determined for other reactions. Some typical values of ρ and σ are given in Table 2.6.

TABLE 2.6
SUBSTITUENT AND REACTION CONSTANTS OF THE HAMMETT
RELATIONSHIP FOR SOME REACTIONS

Substituent Group	$\sigma(\text{meta})$	$\sigma(\text{para})$
CH_3	−0.07	−0.17
C_2H_5	−0.04	−0.15
OH	0.00	−0.46
Cl	0.37	0.23
NO_2	0.71	0.78

Reaction Series	ρ
Benzoic acids ionization (aq), equilibrium	1.000
Phenols ionization (aq), equilibrium	2.113
Benzoylation of aromatic amines in benzene, rate	−2.781

The Hammett correlation runs into difficulty with ortho-substituted compounds because steric hindrance is often of importance in their reactions, a factor not accounted for in linear free-energy relationships. Similarly, aliphatic compounds do not correlate very well. However, a very similar correlative procedure for such compounds has been developed by Taft [R. W. Taft, Jr., *J. Amer. Chem. Soc.*, **74**, 2729, 3120 (1952); **75**, 4231 (1953)], which we may write as

$$\log k' = \log k'_0 + \sigma^* \rho^* \tag{2-125}$$

where k' is a rate or equilibrium constant for a particular member of a reaction series, k'_0 normally the value for the parent methyl compound, ρ^* the reaction constant, and σ^* a constant indicative of the electron attracting capacity of the substituent. While the correlation is similar in form to that of Hammett, parameters ρ^* and σ^* are defined on a different basis. We shall encounter a similar type of correlation for surface reactions in chapter 3.

EXERCISES FOR CHAPTER 2

Section 2.1

1. Compare the fraction, $P(c)$, of hydrogen molecules having a speed of 1 km/sec at 30°C to that of oxygen molecules at the same temperature. Is the ratio of these two fractions the same at 100°C?

2. For the two gases of problem 1, is the energy distribution ratio the same as the speed distribution ratio?

3. Compare the binary collision numbers for hydrogen gas and oxygen gas at 100°C and 1 atm total pressure. What is the ratio of the mean collision times for the two gases under these conditions? What is the collision number for an equimolar mixture of hydrogen and oxygen at 400°C and 1 atm total pressure?

4. Consider the problem of determining the collision number between the molecules of a homogeneous gas and a solid surface. This is equivalent to calculating the rate of effusion of molecules contained in a vessel through an orifice sufficiently small that the effusion rate does not disturb the velocity distribution of molecules remaining in the vessel. For a gas of number density n_A and average speed $\bar{c}$, the geometry may be visualized as shown on Figure 2.15a, where the distance from the orifice of area ds is $\bar{c}$, and the angles θ and ϕ represent its direc-

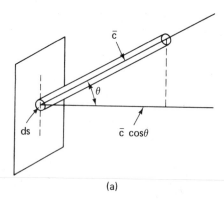

(a)

Figure 2.15 Geometry of the molecular effusion problem (after W. J. Moore, *Physical Chemistry*, 2nd ed., 1955; reprinted by permission of Prentice-Hall, Inc., Englewood Cliffs, N.J.)

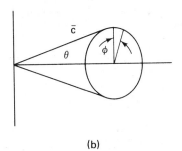

(b)

Figure 2.15 (Cont.)

tion. The differential solid angle containing molecules of speed between $\bar{c}$ and $\bar{c} + d\bar{c}$ at positions between θ and $\theta + d\theta$ and $\phi + d\phi$ is, in polar coordinates (Figure 2.25b), $\sin \theta \, d\theta \, d\phi$. The fraction of the total number of molecules is given by $(\sin \theta \, d\theta \, d\phi / 4\pi)$, where 4π is the total solid angle subtended by the spherical surface of radius $\bar{c}$. From this information, derive equation (2-23).

5. The molecular diameter of oxygen is variously reported as the following:

Diameter ($\mathring{A}$)	Method of determination
2.96	Gas viscosity
2.90	Van der Waals constant
2.34	Molecular refraction
3.75	Closest packing

How much difference is there in the minimum and maximum estimates of collision number at 100°C and 1 atm corresponding to these values?

6. Using the collision-number result for a diameter of 3.75 Å in problem 5, and assuming a collision time of 10^{-13} sec for O_2 molecules and a molecular diameter of 4 Å for O_4 complexes, calculate the number of triple collisions in O_2 at 100°C and 1 atm. What conclusion is one tempted to draw from this concerning the importance of third-order processes?

Section 2.2

7. The decomposition of nitrogen dioxide:

$$2NO_2 \longrightarrow 2NO + O_2$$

obeys second-order kinetics. The following rate-constant data were reported by Bodenstein [M. Bodenstein, *Z. Physik. Chem.*, **100**, 106 (1922)]:

T (°K)	592	603.2	627	651.5	656
k (cm³/mole-sec)	522	755	1,700	4,020	5,030

Using the appropriate collision-theory rate constant, determine what the effective collision diameter is for this reaction based on the data above. Is this quantity a function of temperature?

8. In a binary collision between hard-sphere molecules, what ratio of molecular masses results in the maximum transfer of translational energy?

9. What is $(\Delta E_A/E_A)$ for collision between hydrogen (A) and oxygen (B), each with a line-of-centers velocity equal to one-half the average speed at 25°C given in Table 2.2b?

10. (a) The kinetics of gas-phase isomerization of cyclopropane have been studied by Chambers and Kistakowsky [T. C. Chambers and G. B. Kistakowsky, *J. Amer. Chem. Soc.*, **56**, 399 (1934)], who report the limiting high-pressure rate constant k_∞ to be represented by an Arrhenius form

$$k_\infty = k_\infty^\circ e^{-E/RT}$$

where

$$\log k_\infty^\circ = 15.17 \quad (k_\infty^\circ \text{ in sec}^{-1})$$

$$E = 65.0 \text{ kcal/mole}$$

In terms of the Lindemann theory, what must the effective collision number (k_1) be if the calculation is to agree with the experimental results shown in Figure 2.10? Use $(k^1/k_\infty) = 0.525$ at $\log P$ (cm) $= 0$ as a basis. The temperature is 492°C.

(b) From this value, compute the dependence of (k^1/k_∞) over the pressure range shown in Figure 2.10 and compare with the experimental curve.

11. What energy corresponds to the maximum of the $P(E)$ distribution? What energy corresponds to the average of the $P(E)$ distribution?

12. A certain molecule is nonlinear and contains six atoms. At 400°C, what fraction of these molecules will have an energy of at least 20 kcal/mole? How much does this fraction increase if the temperature is 700°C? What are the corresponding figures for a linear six-atom molecule?

13. For a specified energy level and comparable values of $\bar{\nu}$, what can one conclude concerning the mean lifetimes of reacting molecules as the complexity of the molecule increases?

14. Using collision theory, estimate a rate constant for the recombination of oxygen atoms at 200°C, with $\sigma_{OO} = 3$ Å.

15. Write a detailed kinetic sequence describing all the individual processes involved in the RRK theory for decomposition reactions. From this derive the general expression for the decomposition rate.

Section 2.3

16. Formulate the expression for the rate constant of the reaction

$$H + H_2 \rightleftharpoons H_3^\ddagger \longrightarrow H_2 + H$$

in terms of the appropriate detailed partition functions. Assume that the transition state is linear and symmetric and ignore electronic contributions. Compute the frequency factor for this reaction at 300°K. The moment of inertia of H_2 is 0.459×10^{-40} g-cm^2 and its fundamental vibrational frequency corresponds to a wave number of 4395.2 cm^{-1}, while the moment of inertia of the $H_3^\ddagger$ is estimated to be 3.34×10^{-40} g-cm^2 and its fundamental frequencies are stretching at 3650 cm^{-1}, and doubly degenerate bending at 670 cm^{-1} [H. Eyring and M. Polanyi, *Z. Physik. Chem.*, **B12**, 279 (1931)].

17. The rate constant for the reaction of problem 16 has been determined experimentally by Farkas and Farkas [A. Farkas and L. Farkas, *Proc. Roy. Soc.* (*London*), **A152**, 124 (1935)]. Their result is

$$k = 10^{8.94} T^{1/2} e^{-5500/RT} \text{ liters/mole-sec}$$

Compare this experimental value for frequency factor with that computed in problem 16.

18. The specific rate constant for the homogeneous gas-phase reaction

$$C_2H_4 + H_2 \longrightarrow C_2H_6$$

has the following values:

$$T = 500°K \qquad k = 6.98 \times 10^{-9} \text{ liter/mole-sec}$$
$$T = 600°K \qquad k = 4.16 \times 10^{-6} \text{ liter/mole-sec}$$

At a temperature of 500°K, what are?
(a) The experimental activation energy.
(b) The enthalpy and entropy of activation.
(c) The concentration of transition-state complexes in equilibrium with C_2H_4 and H_2, each at a concentration of 0.05 mole/liter. Assume a linear transition complex.
(d) The steric factor for this reaction.

19. (a) Using the collision-theory expression for the rate constant of a bimolecular reaction, equation (2-33a), and the thermodynamic form of the TST result, from equation (2-108) (in molecular units), show that, approximately,

$$e^{(\Delta S°)^{\ddagger}/R} = p$$

Use order-of-magnitude values for the parameters involved, and a standard state of 1 mole/cm^3. Note that this result provides a simple means for estimating entropies of activation from experimental measurements of steric factors.

(b) For unimolecular reactions, using the approximation that $(kT/h) \approx 10^{13}$ sec^{-1} and $(E/RT) \approx (\Delta H°)^{\ddagger}/RT$, show that

$$(\Delta S°)^{\ddagger} = R \ln (k° \times 10^{-13})$$

where $k°$ is the experimentally determined preexponential factor of an Arrhenius correlation.

20. Consider the simple model for a reaction between two ions in solution given in Figure 2.16. The reactants are conducting spheres of radii r_A and r_B with charge $z_A e$ and $z_B e$, e the electronic charge, and z_A and z_B whole numbers corresponding to the ionic charges. Initial and transition states are as shown on the figure. At a distance x apart the force between them is

$$f = \frac{z_A z_B e^2}{\epsilon x^2}$$

where ϵ is the dielectric constant of the medium. Determine the electrostatic contribution to the free energy of activation. Show that the rate constant for reaction between ions in solution is log-linear with respect to the reciprocal of the dielectric constant.

Solution dielectric constant = ϵ

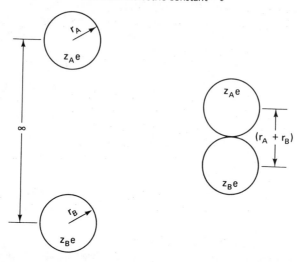

Figure 2.16 Transition-state model for ionic reaction in solution

21. (a) From the concepts of transition-state theory and the thermodynamic relationship

$$V = \left(\frac{\partial G}{\partial P}\right)_T$$

derive an expression for correlating the effect of pressure on the rate constant of a chemical reaction. This will be expressed in terms of an activation volume, $(\Delta V)^{\ddagger}$, corresponding to the free energy of activation, $(\Delta G)^{\ddagger}$.

(b) Determine the value of the activation volume for the hydrolysis of ethyl acetate from the following data:

Pressure (atm)	k (liters/mole-sec)	T (°C)
1	0.080	25
272	0.089	25
544	0.098	25
816	0.107	25

NOTATION FOR CHAPTER 2

Section 2.1

c, c_i	speed (of i), length/time
c_r	mean relative speed, length/time
$\bar{c}, c_m, (\bar{c}^2)^{1/2}$	average, median, root mean square speeds, length/time

$$A + S \rightleftharpoons A \cdot S$$
$$A \cdot S \rightleftharpoons B \cdot S \qquad \text{(XVIII)}$$
$$B \cdot S \rightleftharpoons B + S$$

The first step in this sequence, which we have written as a chemical reaction, represents the adsorption of A on the surface, the second is the surface reaction of A $\cdot$ S to B $\cdot$ S, and the third the desorption of product B from the surface. We will develop our ideas concerning surface reaction by discussing each of these individually. The presentation is based on gas/solid reactions, which are the most common type of heterogeneous catalytic reactions; however, the analysis is also valid for liquid/solid systems. Normally for gas/solid systems the expressions for adsorption equilibrium and reaction rate are written in terms of partial pressures of reactant and product species, whereas in liquid/solid systems concentrations are employed.

3.1   Adsorption and Desorption

a. Ideal surfaces

According to the first step of reaction (XVIII) given above, an essential feature of catalysis is the adsorption of the reacting species on the active surface prior to reaction. Also as indicated, this type of adsorption is generally a very specific interaction between surface and adsorbate which is a chemical reaction in itself and is termed *chemisorption*. *Desorption* is just the reverse of this process, so it is logical to discuss the two together.

To begin with, let us consider the rate at which A molecules in a homogeneous gas will strike a solid surface, which we have seen to be

$$\bar{Z}_{cT} \text{ (surface)} = \tfrac{1}{4} n_A \left(\frac{8kT}{\pi m_A}\right)^{1/2} \qquad \text{(2-23)}$$

This would give us the maximum possible rate of adsorption in any system, if every molecule striking the surface is adsorbed. Now just as in collision theory, to proceed further we must decide upon a more detailed model of what we mean by adsorption, which requires in the present instance more detail on the nature of the surface. The simplest model of a surface we can envision is one in which each adsorption site has the same energy of interaction with the adsorbate, which is not affected by the presence or absence of sorbate molecules on adjoining sites, and in which each site can accommodate only one adsorbate molecule or atom. We might represent the energy contours of such a surface qualitatively as shown in Figure 3.1. Adsorption would occur when a molecule or atom of adsorbate with the required energy strikes

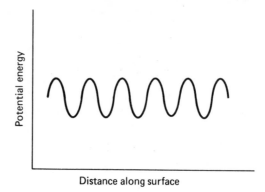

Figure 3.1 Representation of an energetically homogeneous surface

an unoccupied site, and the energy contours would be unaffected by the extent of adsorption.

These requirements for reaction (adsorption) to occur look very similar to those which we imposed on the bimolecular collision number in order to derive the reactive collision number, $\bar{Z}_{\eta}(A, B)$, of equation (2-31). If we designate E as the activation energy required for chemisorption and $C_s = n_s \cdot f(\theta)$, where n_s is the site density per area of surface [the exact nature of $f(\theta)$ need not be specified at this point other than it is a function of the fraction of surface covered by adsorbate, θ_A], as the concentration of sites available for chemisorption, then equation (2-23) may be modified to a form entirely analogous to equation (2-31):

$$\bar{Z}_{cT}(\text{ads}) = \tfrac{1}{4}n_A \left(\frac{8kT}{\pi m_A}\right)^{1/2} n_s \cdot \frac{f(\theta)}{n_s} e^{-E/kT} \tag{3-1}$$

Expressing n_A in terms of the ideal gas law, $n_A = P_A/kT$:

$$\bar{Z}_{cT}(\text{ads}) = \frac{P_A}{\sqrt{2\pi m_A kT}} n_s \cdot \frac{f(\theta)}{n_s} e^{-E/kT} \tag{3-2}$$

We may also include a term analogous to the steric factor to be used as a measure of the deviation of chemisorption rates from this ideal limit:

$$\bar{Z}_{cT}(\text{ads}) = \frac{\sigma P_A}{\sqrt{2\pi m_A kT}} n_s \cdot \frac{f(\theta)}{n_s} e^{-E/kT} \tag{3-3}$$

where σ is such a factor and is commonly termed the *sticking probability*. The adsorption rate constant, k_a°, is thus seen to be $\sigma/\sqrt{2\pi m_A kT}$. Estimates of k_a° or σ for various models of the chemisorption process may be made using TST in a manner similar to that employed for the steric-factor calculations reported in Table 2.4. Some problems of this sort are included in the exercises.

A potential-energy diagram for the adsorption–desorption process $A + S \rightleftharpoons A \cdot S$ is shown in Figure 3.2. As illustrated, the chemisorption is seen to be exothermic, which is in general the case. Also, since adsorption results in a more ordered state, we can argue in thermodynamic terms that entropy changes on chemisorption are negative; this fact will be of use later in testing the reasonableness of rate expressions pertaining to reactions on surfaces.

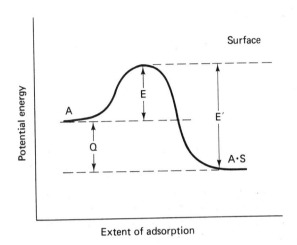

Figure 3.2 Potential-energy diagram for adsorption–desorption

The development of an expression for desorption rates can be made along somewhat more qualitative lines. The rate will be proportional to the concentration of adsorbed species on the surface, C_{AS}, which we will express as $f'(\theta) \cdot n_s$, and will require the activation energy E', as shown in Figure 3.2. Thus, a reasonable form for the rate of desorption is

$$\bar{Z}(\text{des}) = k_d^\circ n_s \cdot \frac{f'(\theta)}{n_s} e^{-E'/kT} \qquad (3\text{-}4)$$

where k_d° is a desorption rate constant per unit surface area. Here again, some more quantitative arguments as to the nature of k_d° can be made in terms of TST, as shown in the problems.

A most important step in the consideration of surface reactions is the equilibrium level of adsorption on a surface. In terms of the example $A + S \rightleftharpoons A \cdot S$, at this point the rate of adsorption will equal the rate of desorption; as in any chemical change, the process is one of dynamic equilibrium, where the net rate of change is zero. From equations (3-3) and (3-4) we have

$$\frac{\sigma P_A}{\sqrt{2\pi m_A kT}} n_s \cdot f(\theta) e^{-E/kT} = k_d^\circ n_s \cdot f'(\theta) e^{-E'/kT} \qquad (3\text{-}5)$$

This expression is extraordinarily useful, since it permits us to obtain a relationship between the surface coverage factors, $f(\theta)$ and $f'(\theta)$, which cannot be conveniently measured, and macroscopic quantities such as P_A and the heat of adsorption, which can. If we solve equation (3-5) for the ratio $f'(\theta)/f(\theta)$,

$$\frac{f'(\theta)}{f(\theta)} = \frac{\sigma P_A e^{Q/kT}}{k_d^\circ \sqrt{2\pi m_A kT}} \tag{3-6}$$

it can be seen that the surface coverage at equilibrium (or at least some function of it) is determined by the temperature of the system and the partial pressure of adsorbate. Such an equation for fixed temperature and varying partial pressure expresses the *adsorption isotherm* for the adsorbate, or for fixed partial pressure and varying temperature the *adsorption isobar*. The heat of adsorption, Q, appears in equation (3-6) since in solving for the ratio $f'(\theta)/f(\theta)$ the difference $(E' - E)$ appears in the exponential; from Figure 3.2 we see that this is equal in magnitude to the heat of adsorption.

Equation (3-6) is a form of the Langmuir isotherm [I. Langmuir, *J. Amer. Chem. Soc.*, **40**, 1361 (1918)], which we may write in more general terms as

$$\frac{f'(\theta)}{f(\theta)} = K_A P_A = K_A^\circ e^{Q/kT} P_A \tag{3-6a}$$

where

$$K_A = \frac{\sigma e^{Q/kT}}{k_d^\circ \sqrt{2\pi m_A kT}}$$

Specific forms of this equation will depend on what $f'(\theta)$ and $f(\theta)$ are. For the example we have been using, the surface concentration of sites available for chemisorption, C_S, is obviously equal to the total number of sites per area times the fraction which are not occupied by A:

$$C_S = n_s(1 - \theta_A) \tag{3-7}$$

where θ_A is the fraction of sites occupied by A. Similarly, the surface concentration of adsorbate, C_{AS}, is

$$C_{AS} = n_s\theta_A \tag{3-8}$$

so that

$$f'(\theta) = \theta_A$$
$$f(\theta) = 1 - \theta_A \tag{3-9}$$

Inserting these values into equation (3-6a) and solving for θ_A gives

$$\theta_A = \frac{K_A P_A}{1 + K_A P_A} \tag{3-10}$$

which is the Langmuir isotherm equation for the nondissociative (i.e., single molecule on single site) adsorption of a single species on a surface.

In many cases of practical importance, the adsorbate molecule will dissociate on adsorption [H_2 on many metals; see, for example, J. K. Roberts, *Proc. Roy. Soc. (London)*, **A152**, 445 (1935); O. Beeck and A. W. Ritchie, *Disc. Faraday Soc.*, **8**, 159 (1950)] or occupy two sites by bonding at two points in the molecule [ethylene on nickel; G. H. Twigg and E. K. Rideal, *Proc. Roy. Soc. (London)*, **A171**, 55 (1939)]. In such cases,[1]

$$f(\theta_A) = (1 - \theta_A)^2$$
$$f'(\theta_A) = \theta_A^2 \tag{3-11}$$

and the corresponding isotherm equation is

$$\theta_A = \frac{(K_A P_A)^{1/2}}{1 + (K_A P_A)^{1/2}} \tag{3-12}$$

A second modification of practical importance is when more than one adsorbate species is on the surface. For example, consider the equilibrium (no surface reaction)

$$A + S \rightleftharpoons A \cdot S$$
$$B + S \rightleftharpoons B \cdot S \tag{XIX}$$

where A and B are chemisorbed on the same surface site, S; that is, they are competitively adsorbed on the surface. The reader will verify in problem 5 that the corresponding isotherm equation for the surface coverages A and B, θ_A and θ_B, respectively, are

$$\theta_A = \frac{K_A P_A}{1 + K_A P_A + K_B P_B}$$
$$\theta_B = \frac{K_B P_B}{1 + K_A P_A + K_B P_B} \tag{3-13}$$

A major property of the Langmuir isotherm is that of saturation. In equation (3-10), for example, when $K_A P_A \gg 1$, $\theta_A \to 1$, and no further adsorption occurs. This is a result of the surface model in which each adsorption site can accommodate only one adsorbate molecule. Saturation of the surface, then, corresponds to the occupancy of all sites and is normally referred to as *monolayer coverage*. At low pressures, $K_A P_A \ll 1$ and equation (3-10) assumes a linear form in P_A corresponding to Henry's law adsorption. These general features of a Langmuir isotherm are shown in Figure 3.3. Experimentally, one measures either the weight or volume of material adsorbed, and the ratio of this quantity at a given partial pressure to that at saturation can be taken as a direct measure of the surface coverage. This is indicated in the figure, where $\theta_A = V/V_M$, in which V is the volume adsorbed and V_M the volume adsorbed at saturation.

The interpretation of data on adsorption in terms of the Langmuir isotherm is most easily accomplished using the procedure described previously for reaction rate data (i.e., examination of results in terms of a linear form

[1] If the adsorbate molecule is immobile on the surface, the occupancy of nearest-neighbor sites must be taken into account [A. R. Miller, *Proc. Cambridge Phil. Soc.*, **43**, 232 (1947)]. This is a refinement, however, beyond the scope of the present treatment and equations (3-11) will be employed without reference to the detailed nature of the adsorbate layer.

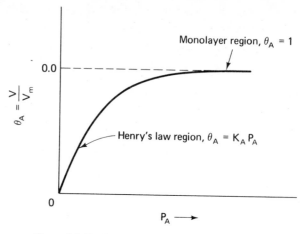

Figure 3.3 Typical Langmuir adsorption isotherm

of the isotherm equation). Rearrange equation (3-10) to

$$\frac{1}{V} = \left(\frac{1}{P_A}\right)\left(\frac{1}{K_A V_M}\right) + \frac{1}{V_M} \tag{3-14}$$

which form indicates that a plot of $(1/V)$ versus $(1/P_A)$ should be linear with slope $(1/K_A V_M)$ and intercept $(1/V_M)$.[1] Figure 3.4 presents a representation of some results for adsorption of hydrogen on copper and a corresponding

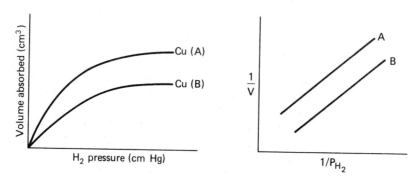

Figure 3.4 Langmuir interpretation of H_2 adsorption on copper powder (after A. F. H. Ward, *Proc. Roy. Soc.* (*London*), **A133**, 506 (1931); with permission of The Royal Society)

[1] In many sources it is recommended to plot according to the alternative linear form:

$$\left(\frac{P_A}{V}\right) = \frac{P_A}{V_M} + \frac{1}{K_A V_M}$$

[i.e., (P_A/V) versus P_A]. This has the disadvantage of involving P_A on both axes and may involve some loss in sensitivity of the test. "Seek simplicity and distrust it" (A. N. White-head).

Langmuir interpretation which provides a satisfactory correlation of the data. Examples of false obedience to the Langmuir isotherm are numerous. These usually arise when adsorption data have not been obtained over a sufficiently wide range of partial pressures; a good test is to see if values of V_M evaluated from isotherm data at different temperatures are equal, since within the framework of the Langmuir surface model, the saturation capacity should not vary with temperature. See problem 6.

b. Nonideal surfaces

It is apparent that no real surface could have the potential-energy contour depicted in Figure 3.1, nor is it reasonable to expect that adsorbate molecules on the surface would not interact with each other. In both events, one would expect to observe some distribution of energies of interaction with the surface which might be correlated with the extent of surface coverage. Since the strongest interactions would occur at the beginning of adsorption, the experimental observation would be that of a decreasing heat of chemisorption with increasing coverage. It is not possible to distinguish from this information alone whether energetic inhomogeniety of the surface or adsorbate interactions are the cause, but in most of the practical applications with which we are concerned, this amount of detail is not necessary. In the following discussion we shall consider the case of an inhomogeneous surface, using the specific case of nondissociative adsorption of a single species for illustration.

Consider that the surface may be divided into a number of groups of sites, each with similar heats of adsorption and so capable of being represented by a Langmuir isotherm. The fractional coverage of each of these for a single sorbate A is θ_{A_i}:

$$\theta_{A_i} = \frac{K_{A_i} P_A}{1 + K_{A_i} P_A} \qquad (3\text{-}10a)$$

and the total coverage is

$$\theta_A = \sum_i \theta_{A_i} n_i \qquad (3\text{-}15)$$

where n_i is the fraction of the total number of adsorption sites belonging to group i and can be represented by an appropriate distribution function. It is reasonable to write this distribution function in terms of the heat of chemisorption, Q, and in one of the major isotherm expressions for nonideal surfaces, the *Freundlich isotherm*, this distribution is taken to be an exponential one. That is,

$$dn_i = n_0 e^{-Q/Q_M} dQ \qquad (3\text{-}16)$$

where n_0 and Q_M are scaling constants and dn_i the fraction of sites with energies between Q and $Q + dQ$. If we further assume, as indicated by the form of equation (3-16), that the groups of sites do not differ greatly in energy level, the summation may be replaced by integration with respect to the distribution function:

$$\theta_A = \int_0^\infty n_0 e^{-Q/Q_M} \frac{K_{A_t} P_A}{1 + K_{A_t} P_A} \, dQ \qquad (3\text{-}17)$$

Recalling from equation (3-6a) that K_{A_t} is also a function of Q, we have:

$$\theta_A = \int_0^\infty n_0 e^{-Q/Q_M} \frac{K_A^\circ e^{Q/kT} P_A}{1 + K_A^\circ e^{Q/kT} P_A} \, dQ \qquad (3\text{-}18)$$

which, for $Q \gg kT$ can be integrated to

$$\theta_A = (K_A^\circ P_A)^{kT/Q_M} n_0 Q_M \qquad (3\text{-}19)$$

It can be shown that the product of the scaling constants, $n_0 Q_M$, is unity, so that

$$\theta_A = (K_A^\circ P_A)^{kT/Q_M} \qquad (3\text{-}20)$$

This is the basic Freundlich isotherm, which is a power law of the form

$$\theta = cP^m \qquad (3\text{-}20a)$$

The exponential distribution law employed defines the nature of the change in heat of chemisorption with surface coverage. The *Clausius–Clapeyron equation* defines the heat effect accompanying a change of phase, according to which

$$-Q = k\left[\frac{d \ln P_A}{d(1/T)}\right]_{\theta_A} \qquad (3\text{-}21)$$

Substituting from the isotherm equation and carrying out the indicated differentiation gives

$$Q = -Q_M \ln \theta_A \qquad (3\text{-}22)$$

so that the heat of adsorption is logarithmically dependent on surface coverage.

Testing experimental data for obedience to the isotherm can be done by a log–log plot of volume adsorbed versus adsorbate partial pressure, in accordance with the linearized form of equation (3-20):

$$\log V = \log V_M + \frac{kT}{Q_M} \log K_A^\circ + \frac{kT}{Q_M} \log P_A \qquad (3\text{-}23)$$

where V and V_M are the volume adsorbed and the monolayer volume, respectively. In Figure 3.5 are given some data for the adsorption of hydrogen on tungsten powder and the corresponding Freundlich interpretation.

A second type of isotherm which has been extensively employed for non-ideal surfaces is that of Temkin, which postulates that the heat of chemisorption decreases linearly with surface coverage:

$$Q = Q_0(1 - \alpha\theta) \qquad (3\text{-}24)$$

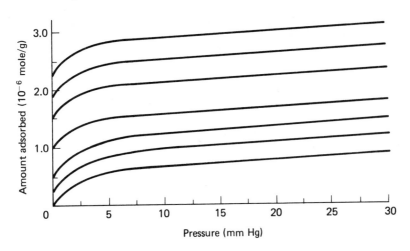

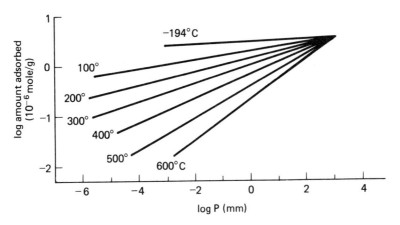

Figure 3.5 Freundlich interpretation of H_2 adsorption on tungsten powder (reprinted with permission from W. G. Frankenburg, *J. Amer. Chem. Soc.*, **66**, 1827 (1944); copyright by the American Chemical Society)

where Q_0 is the heat of chemisorption at zero coverage and α the linear scaling constant. By a derivation similar to that employed for the Freundlich isotherm, we obtain

$$\theta_A = \frac{kT}{Q_0\alpha} \ell n\, K_A^\circ e^{-Q_0/kT} P_A \qquad (3\text{-}25)$$

in which the correlation is limited to surface coverages in the midrange (i.e., $0.25 < \theta_A < 0.75$). This isotherm is perhaps best known for its correlation of the important system nitrogen and hydrogen on iron, involved in ammonia synthesis. For an extensive discussion of isotherm expressions for both ideal and nonideal systems and many experimental examples, the reader is referred to the monograph by Hayward and Trapnell (D. O. Hayward and B. M. W. Trapnell, *Chemisorption*, 2nd ed., Butterworth, Washington, D.C., 1964).

3.2 Surface Reactions with Rate-Controlling Steps

In the introduction to this chapter it was stated that reaction sequences involving surface steps such as (XVIII) could be visualized as a type of chain reaction. This is indeed so, and we shall have more to say concerning the analysis of surface reactions in terms of the pssh a little later. However, particularly associated with the development of the theory of surface reaction kinetics has been the concept of the rate-limiting or rate-controlling step. This presents a rather different view of sequential steps than does chain reaction theory, since if a single step controls the rate of reaction, all other steps must be at equilibrium—a result that is not a consequence of the general pssh.

In fact, pursuing the example of (XVIII) a bit further in this regard, if the surface reaction step is rate-limiting, we can express the net rate of reaction directly in terms of the surface species concentrations, C_{AS} and C_{BS}:

$$\text{rate} = k_{S1}C_{AS} - k_{S2}C_{BS} \qquad (3\text{-}26)$$

where k_{S1} and k_{S2} are velocity constants for the forward and reverse surface reaction steps. This represents a considerable simplification from the normal chain reaction analysis, because the first and third steps of (XVIII) are at equilibrium and the surface species concentrations are entirely determined by their adsorption/desorption equilibrium on the surface. If we write, as previously, the surface concentrations in terms of surface coverage and total site density:

$$C_{AS} = n_s \cdot \theta_A$$
$$C_{BS} = n_s \cdot \theta_B \qquad (3\text{-}27)$$

Substitution of equations (3-13) and (3-27) into (3-26) gives the rate of reaction in terms of the partial pressures of reacting species:

$$\text{rate} = \frac{k_{S1}n_s K_A P_A}{1 + K_A P_A + K_B P_B} - \frac{k_{S2}n_s K_B P_B}{1 + K_A P_A + K_B P_B}$$

or

$$\qquad (3\text{-}28)$$

$$\text{rate} = \frac{k_{S1}n_s K_A P_A - k_{S2}n_s K_B P_B}{1 + K_A P_A + K_B P_B}$$

In fact, the value of n_s may be a somewhat elusive quantity, so that in practice this is absorbed into the rate constant as, say, $k'_{S1} = (k_{S1}n_s)$.

The rates of catalytic reactions are most often expressed in terms of unit mass or unit total surface (not external surface) of the catalyst. Thus, an appropriate working form of equation (3-28) for the rate of disappearance of A in (XVIII) is

$$-\frac{1}{W_C}\frac{dN_A}{dt} = \frac{k'_{S1}K_AP_A - k'_{S2}K_BP_B}{1 + K_AP_A + K_BP_B} = -r_A \qquad (3\text{-}28a)$$

or

$$-\frac{1}{A_C}\frac{dN_A}{dt} = \frac{k'_{S1}K_A{}^PA - k'_{S2}K_BP_B}{1 + K_AP_A + K_BP_B} = -r_A \qquad (3\text{-}28b)$$

where W_C is the weight of catalyst, A_C the total surface area,[1] and the specific rate constants k'_{S1} and k'_{S2} defined accordingly as per unit mass or area. Since there is a wide variation in surface area per unit mass in various catalysts, rate definitions based on mass can be misleading as to the true activity of a catalyst. If information on A_C is available, specific area rates are preferable.

As a second example of surface-reaction-rate control, consider the slightly more complicated case of a bimolecular reaction occurring on the same sites, S, of a surface. Overall, $A + B \rightleftharpoons C + D$, and the individual steps are:

$$A + S \rightleftharpoons A \cdot S$$
$$B + S \rightleftharpoons B \cdot S$$
$$A \cdot S + B \cdot S \rightleftharpoons C \cdot S + D \cdot S \qquad (k_{S1}, k_{S2}) \qquad (XX)$$
$$C \cdot S \rightleftharpoons C + S$$
$$D \cdot S \rightleftharpoons D + S$$

and the rate of reaction of A is

$$-r_A = k_{S1}C_{AS}C_{BS} - k_{S2}C_{CS}C_{DS} \qquad (3\text{-}29)$$

The appropriate form of equation (3-13) for the competitive adsorption of A, B, C, and D on the surface is

$$\theta_i = \frac{K_iP_i}{1 + K_AP_A + K_BP_B + K_CP_C + K_DP_D} \qquad (i = A, B, C, D)$$

and the rate equation is

$$-r_A = \frac{k'_{S1}K_AK_BP_AP_B - k'_{S2}K_CK_DP_CP_D}{(1 + K_AP_A + K_BP_B + K_CP_C + K_DP_D)^2} \qquad (3\text{-}30)$$

where the K's represent the adsorption equilibrium constants for the individual species on the surface.

It is clear from equations (3-28) and (3-30) that when a surface reaction

---

[1]The characterization of catalysts in terms of total surface area, porosity, pore structure, and so on, is a problem of considerable magnitude in itself. We shall not attempt to define the precise nature or determination of A_C at this point, but will treat several aspects of catalyst characterization together in chapter 7.

step is rate-controlling, one needs only information concerning adsorption equilibria of reactant and product species in order to write the appropriate rate equation for the overall transformation. It also often occurs that the products of a reaction are much less strongly bound to the surface than are the reactants, or that the surface reaction step is essentially irreversible. In such cases the product terms disappear from equations such as (3-28) or (3-30), and in the older literature this overall process is referred to as a *Langmuir–Hinshelwood* (LH) *mechanism*. More commonly now, however, equations of the general form of (3-28) or (3-30), referring to surface-reaction-steps rate control, are termed *Langmuir–Hinshelwood rate equations*. A bewildering array of such equations may be derived for various types of reactions, depending on specific assumptions concerning the magnitude of adsorption equilibrium constants, dissociative or nondissociative adsorption of species, type of sites involved in adsorption and surface reaction, and so on. Some examples of these are given in Table 3.1. In each of these the active centers or sites form a closed sequence in the overall reaction. According to this view, when the stoichiometry between reactants and products is not balanced, the concentration of vacant active centers also enters the rate equation [reaction (2) Table 3.1]. In formulation of the overall rate equation, this concentration may be written in terms of occupied active centers by the relationship

$$C_S = \frac{n_s}{1 + \sum_i K_i P_i} \tag{3-31}$$

where the summation is with respect to all species adsorbed on the surface. The denominator of all LH rate equations consists of a summation of the adsorption terms for species on the surface; these are sometimes called *adsorption inhibition terms* (particularly with reference to product adsorption) since they decrease the magnitude of the rate as given by the power-law functions that make up the numerator of LH equations.

The bimolecular reaction involving one reactant nonadsorbed [reaction (6), Table 3.1] is sometimes referred to as a *Rideal* or *Eley–Rideal mechanism*. It has been used primarily in interpretation of catalytic hydrogenation kinetics.

The analysis of rate-controlling steps other than surface reaction on ideal surfaces has been developed in detail by Hougen and Watson [O. A. Hougen and K. M. Watson, *Chemical Process Principles*, vol. III, Wiley, New York, 1947; see also K. H. Yang and O. A. Hougen, *Chem. Eng. Progr.*, **46**, 146 (1950)]. If, for example, the rate of adsorption of a reactant species on the surface is slow compared to other steps, it is no longer correct to suppose that its surface concentration is determined by adsorption equilibrium. Rather, there must be chemical equilibrium between all species on the surface, and the surface concentration of the species involved in the rate-limiting step is determined by this equilibrium. The actual partial pressure of the rate-limit-

TABLE 3.1
SOME EXAMPLES OF LANGMUIR–HINSHELWOOD RATE EQUATIONS

Reaction Description	Individual Steps	Equation, $(-r_A)$	Constants
(1) Isomerization: $A \rightleftharpoons B$	$A + S \rightleftharpoons A \cdot S$ $A \cdot S \rightleftharpoons B \cdot S$ $B \cdot S \rightleftharpoons B + S$	$-r_A = k_{S1} C_{AS} - k_{S2} C_{BS}$ $= \dfrac{k'_{S1} K_A P_A - k'_{S2} K_B P_B}{1 + K_A P_A + K_B P_B}$	$k'_{S1} = k_{S1} n_s$ $k'_{S2} = k_{S2} n_s$
(2) Decomposition: $A \rightleftharpoons B + C$	$A + S \rightleftharpoons A \cdot S$ $A \cdot S + S \rightleftharpoons B \cdot S + C \cdot S$ $B \cdot S \rightleftharpoons B + S$ $C \cdot S \rightleftharpoons C + S$	$-r_A = k_{S1} C_{AS} C_S - k_{S2} C_{BS} C_{CS}$ $= \dfrac{k'_{S1} K_A P_A - k'_{S2} K_B K_C P_B P_C}{(1 + K_A P_A + K_B P_B + K_C P_C)^2}$	$k'_{S1} = k_{S2} n_s^2$ $k'_{S2} = k_{S2} n_s^2$ $C_S = \dfrac{n_s}{(1 + K_A P_A + K_B P_B + K_C P_C)}$
(3) Bimolecular: $A + B \rightleftharpoons C + D$	$A + S \rightleftharpoons A \cdot S$ $B + S \rightleftharpoons B \cdot S$ $A \cdot S + B \cdot S \rightleftharpoons C \cdot S + D \cdot S$ $C \cdot S \rightleftharpoons C + S$ $D \cdot S \rightleftharpoons D + S$	$-r_A = k_{S1} C_{AS} C_{BS} - k_{S2} C_{CS} C_{DS}$ $= \dfrac{k'_{S1} K_A K_B P_A P_B - k'_{S2} K_C K_D P_C P_D}{(1 + K_A P_A + K_B P_B + K_C P_C + K_D P_D)^2}$	$k'_{S1} = k_{S1} n_s^2$ $k'_{S2} = k_{S2} n_s^2$
(4) Bimolecular, different sites: $A + B \rightleftharpoons C + D$	$A + S_1 \rightleftharpoons A \cdot S_1$ $B + S_2 \rightleftharpoons B \cdot S_2$ $A \cdot S_1 + B \cdot S_2 \rightleftharpoons C \cdot S_1 + D \cdot S_2$ $C \cdot S_1 \rightleftharpoons C + S_1$ $D \cdot S_2 \rightleftharpoons D + S_2$	$-r_A = k_{S1} (C_{AS})_1 (C_{BS})_2 - k_{S2} (C_{CS})_1 (C_{DS})_2$ $= \dfrac{k'_{S1} (K_A)_1 (K_B)_2 P_A P_B - k'_{S2} (K_C)_1 (K_D)_2 P_C P_D}{[1 + (K_A)_1 P_A + (K_C)_1 P_C][1 + (K_B)_2 P_B + (K_D)_2 P_D]}$	$k'_{S1} = k_{S1} n_{s1} n_{s2}$ $k'_{S2} = k_{S2} n_{s1} n_{s2}$
(5) Bimolecular with dissociation of one reactant; $\frac{1}{2} A_2 + B \rightleftharpoons C + D$	$A_2 + 2S \rightleftharpoons 2A \cdot S$ $B + S \rightleftharpoons B \cdot S$ $A \cdot S + B \cdot S \rightleftharpoons C \cdot S + D \cdot S$ $C \cdot S \rightleftharpoons C + S$ $D \cdot S \rightleftharpoons D + S$	$-r_A = k_{S1} C_{AS} C_{BS} - k_{S2} C_{CS} C_{DS}$ $= \dfrac{k'_{S1} (K_A P_A)^{1/2} K_B P_B - k'_{S2} K_C K_D P_C P_D}{[1 + (K_A P_A)^{1/2} + K_B P_B + K_C P_C + K_D P_D]^2}$	$k'_{S1} = k_{S1} n_s^2$ $k'_{S2} = k_{S2} n_s^2$
(6) Bimolecular with one reactant not adsorbed; $A(g) + B \rightleftharpoons C$	$B + S \rightleftharpoons B \cdot S$ $A(g) + B \cdot S \rightleftharpoons C \cdot S$ $C \cdot S \rightleftharpoons C + S$	$-r_A = k_{S1} C_{BS} C_A - k_{S2} C_{CS}$ $= \dfrac{k'_{S1} K_B P_B (P_A / RT) - k'_{S2} K_C P_C}{1 - K_B P_B + K_C P_C}$	$k'_{S1} = k_{S1} n_s$ $k'_{S2} = k_{S2} n_s$

ing adsorbate consequently will not appear in the adsorption terms of the rate equation, but is replaced by that pressure corresponding to the surface concentration level established by the equilibrium of other steps. This is sometimes called the *virtual pressure* of the species in question. Again consider the isomerization example of (XVIII), this time in which the rate of adsorption of A is controlling. From equations (3-3) and (3-4) we may write net rate of reaction as the difference between rates of adsorption and desorption of A:

$$-r_A = k_a P_A C_S - k_d C_{AS} \tag{3-32}$$

in which k_a and k_d are adsorption and desorption rate constants (i.e., $k_a = k_a^\circ e^{-E/RT}$) and C_{AS} is the surface concentration of A as determined by the equilibrium of all other steps. If we define a surface reaction equilibrium constant, K_S, in the normal way,

$$K_S = \frac{C_{BS}}{C_{AS}}$$

or

$$C_{AS} = \frac{C_{BS}}{K_S} \tag{3-33}$$

According to the discussion above, we can define C_{AS} also from the normal isotherm equation where the virtual pressure, P_A^*, is used in place of the actual partial pressure, P_A:

$$C_{AS} = n_s \theta_A = \frac{n_s K_A P_A^*}{1 + K_A P_A^* + K_B P_B} \tag{3-34}$$

and similarly for C_{BS}:

$$C_{BS} = n_s \theta_B = \frac{n_s K_B P_B}{1 + K_A P_A^* + K_B P_B} \tag{3-35}$$

Substituting (3-34) and (3-35) into the expression for K_S and solving for P_A^* gives

$$P_A^* = \frac{K_B P_B}{K_A K_S} \tag{3-36}$$

Employing these expressions for C_{AS}, C_{BS}, P_A^*, and equation (3-31) for C_S gives us the final rate equation from (3-32):

$$-r_A = \frac{k_A' P_A - (k_d' K_B P_B / K_S)}{1 + (K_B P_B / K_S) + K_B P_B} \tag{3-37}$$

An entirely analogous procedure may be used for a desorption rate-controlling step in this example, with $(-r_A)$ determined by

$$(-r_A) = k_d C_{BS} - k_a P_B C_S \tag{3-38}$$

with the appropriate surface concentrations defined in terms of the virtual pressure of B. Application of these procedures to more complicated reaction schemes is left to the problems.

a. Arrows[1]

Sometimes it is not altogether clear what is meant or implied by the concept of a rate-limiting step. Indeed, the reader may have his or her own questions at this point, since in the preceding discussion we have purposely written all steps as reversible, indicating some influence of the reverse reaction even in those taken to be rate-limited by the forward reaction. The best way to approach this is by considering separately the forward and reverse rates of the individual steps. Let us take the first two steps of (XVIII) for illustration:

$$A + S \rightleftharpoons A \cdot S$$
$$A \cdot S \rightleftharpoons B \cdot S$$

where the forward rate of the first step is f_1, the reverse rate of that step, f_{-1}, and correspondingly f_2 and f_{-2} for the second step. Now represent the magnitude of these individual rates by the length of the arrows shown in Figure 3.6 for a given situation. The overall rate of reaction, r, positive for product and negative for reactant, is

$$r = f_1 - f_{-1} = \Delta_1$$

and also

$$\Delta_1 = \Delta_2 = f_2 - f_{-2}$$

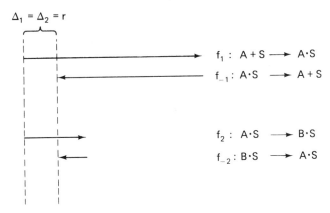

Figure 3.6 Individual forward and reverse rates for an example sequence of chemisorption and surface reaction; rate-determining step

[1]The approach of this section has been developed in considerable detail for a number of problems in the text by Boudart, who attributes its origin to K. Tamaru.

This rate is indicated on the figure by the line-segment length between the vertical dashed lines at the left. Yet, while $\Delta_1 = \Delta_2$, the individual rates f_1 and f_{-1} are much larger than are f_2 and f_{-2}. We can define an extent of approach to equilibrium for these two steps as

$$\frac{f_1 - f_{-1}}{f_1} \quad \text{and} \quad \frac{f_2 - f_{-2}}{f_2} \tag{3-39}$$

in which the limit of zero (i.e., $f_1 = f_{-1}$) defines equilibrium and unity complete irreversibility. It is apparent from the diagram that by this measure the first step is much closer to equilibrium, since the numerators of both expressions are equal, but $f_1 \gg f_2$. In this case one would identify the second step as rate-controlling and the first step as being in at least a near-equilibrium state.

Now suppose, however, that the individual rates appear as in Figure 3.7. Here f_1 is not very much larger than f_2, and by the measure of equation (3-39) one would conclude that both steps are in a state of quasi-equilibrium. The concept of a rate-limiting step breaks down in this case, and it is necessary to turn to more general methods of analysis to obtain the appropriate rate equations.

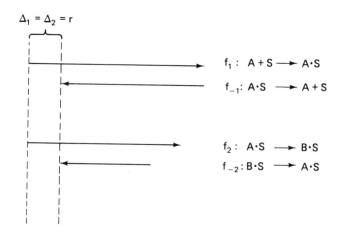

Figure 3.7 Individual forward and reverse rates for an example sequence of chemisorption and surface reaction; no rate-determining step

b. Chains

As stated before, in the most general sense we may treat surface reactions as chain reactions that involve a closed sequence of active centers. Under conditions where the pssh is applicable, we may analyze the individual steps through the time independence of the concentration of each active center and the fact that

$$n_s = C_S + \sum_i C_{S_i}$$ (3-40)

A general method for the solution of the kinetics of certain types of sequences has been given by Christiansen [J. A. Christiansen, *Advan. Catalysis*, **5**, 311 (1953)]. The method may be applied when the rate of each step in the sequence is linear (first order) in a single active center. Using the notation of chapter 1 for chain reactions, we may write a general sequence of this sort as

$$\ldots R_1 \rightleftharpoons \ldots R_2 \qquad (1)$$

$$\vdots$$

$$\ldots R_n \rightleftharpoons \ldots R_{n+1} \qquad (n) \qquad \text{(XXI)}$$

$$\vdots$$

$$\ldots R_N \rightleftharpoons \ldots R_1 \qquad (N)$$

where the Nth step closes the sequence by regeneration of R_1. If, as in chapter 1, we designate a rate for each step as equal to the product of a rate factor containing all rate constants and concentrations except the concentration of active center, then

$$f_1 = r_1 R_1, \quad f_{-1} = r_{-1} R_2, \quad \ldots$$ (3-41)

where $r_1, \ldots$ are the rate factors and $R_1, \ldots$ the active-center concentrations. Now, the net rate in each step must be the same, as indicated in Figures 3.6 and 3.7, so

$$1 = r_1\left(\frac{R_1}{r}\right) - r_{-1}\left(\frac{R_2}{r}\right)$$

$$1 = r_n\left(\frac{R_n}{r}\right) - r_{-n}\left(\frac{R_{n+1}}{r}\right) \qquad (3\text{-}42)$$

$$1 = r_N\left(\frac{R_N}{r}\right) - r_{-N}\left(\frac{R_1}{r}\right)$$

This is a linear set of equations in terms of the active-center concentrations and known coefficients $r_1, \ldots, r_{-N}$. In matrix form,

$$r\frac{R}{r} = 1$$ (3-42a)

for which the solution is

$$\frac{R_i}{r} = \frac{\det(R_i/r)}{\det(R/r)}$$ (3-43)

The determinants of equation (3-43) are given by

$$\det\left(\frac{\mathbf{R}}{r}\right) = \begin{vmatrix} r_1 & -r_{-1} & 0 & 0 & \cdots & 0 \\ 0 & r_2 & -r_{-2} & 0 & \cdots & 0 \\ 0 & 0 & r_3 & -r_{-3} & \cdots & 0 \\ \cdot & & & & & \\ \cdot & & & & & \\ \cdot & & & & & \\ -r_{-N} & \multicolumn{5}{c}{\cdots\cdots\cdots\cdots\cdots\cdots\cdots\cdots\cdots\cdots} & r_N \end{vmatrix}$$

and

$$\det\left(\frac{\mathbf{R}_1}{r}\right) = \begin{vmatrix} 1 & -r_{-1} & 0 & 0 & \cdots & 0 \\ 1 & r_2 & -r_{-2} & 0 & \cdots & 0 \\ 1 & 0 & r_3 & -r_{-3} & \cdots & 0 \\ \cdot & & & & & \\ \cdot & & & & & \\ \cdot & & & & & \\ 1 & \multicolumn{5}{c}{\cdots\cdots\cdots\cdots\cdots\cdots\cdots\cdots\cdots\cdots} & r_N \end{vmatrix}$$

and so on. From equation (3-40),

$$\frac{n_s}{r}\sum_i \frac{\mathbf{R}_i}{r} = \frac{\sum_i \det(\mathbf{R}_i/r)}{\det(\mathbf{R}/r)} \tag{3-44}$$

since $R_1 = C_s$, $R_2 = C_{s1}$, and so on. The rate of reaction is then

$$r = \frac{n_s \det(\mathbf{R}/r)}{\sum_i \det(\mathbf{R}_i/r)} \tag{3-45}$$

For the isomerization reaction, (XVIII), we have been using as an example:

$$\ldots + R_1 \rightleftharpoons R_2$$
$$R_2 \rightleftharpoons R_3 \tag{XVIIIa}$$
$$R_3 \rightleftharpoons R_1 + \ldots$$

and the application of equation (3-45) gives for the rate

$$r = \frac{n_s(r_1 r_2 r_3 - r_{-1} r_{-2} r_{-3})}{r_1 r_2 + r_1 r_3 + r_1 r_{-2} + r_2 r_3 + r_2 r_{-3} + r_3 r_{-1} + r_{-1} r_{-2} + r_{-1} r_{-3} + r_{-2} r_{-3}} \tag{3-46}$$

Now let us examine what form equation (3-46) will assume if we postulate a rate-determining surface reaction step. In this case terms in the denominator containing r_2 or r_{-2} will be small in comparison to those which do not contain them, and:

$$r = \frac{n_s(r_1 r_2 r_3 - r_{-1} r_{-2} r_{-3})}{r_1 r_3 + r_3 r_{-1} + r_{-1} r_{-3}} \tag{3-46a}$$

5. keeping in mind that the representation will be valid only for a limited range of operating conditions.

In Table 3.2 is given a summary of some comparisons provided by Weller on the basis of this argument; the power-law correlations generally provide

TABLE 3.2

COMPARISON OF POWER-LAW AND LANGMUIR–HINSHELWOOD RATE EQUATIONS

I. SO$_2$ Oxidation[a]

LH Form:	Power-Law Form:
$$r = \frac{k(P_{SO_2}P_{O_2}^{1/2} - P_{SO_3}/K)}{[1 + (K_{O_2}P_{O_2})^{1/2} + K_{SO_3}P_{SO_3}]^2}$$	$$r = k[P_{SO_2}P_{SO_3}^{-1/2} - P_{SO_3}^{-1/2}P_{O_2}^{1/2}/K]$$
% deviation = 15.4 (average of 12 experiments on variation of P_{SO_2} and P_{SO_3})	% deviation = 13.3 (average of 12 experiments on variation of P_{SO_2} and P_{SO_3})

II. Hydrogenation of Codimer[b]

LH Form:	Power-Law Form:
$$r = \frac{kP_H P_U}{(1 + K_H P_H + K_U P_U + K_S P_S)^2}$$	$$r = kP_U^{1/2}P_H^{1/2}$$
H = hydrogen U = codimer S = product	
% Deviation: 20.9 (200°C) 19.6 (275°C) 19.4 (325°C)	% Deviation 19.6 (200°C) 32.9 (275°C) 21.4 (325°C)

III. Phosgene Synthesis[c]

LH Form:	Power-Law Form:
$$r = \frac{kK_{CO}K_{Cl_2}P_{CO}P_{Cl_2}}{(1 + K_{Cl_2}P_{Cl_2} + K_{COCl_2}P_{COCl_2})^2}$$	$$r = kP_{CO}P_{Cl_2}^{1/2}$$
% Deviation 3.4 (30.6°C) 5.6 (42.7°C) 2.6 (52.5°C) 7.0 (64.0°C)	% Deviation: 13.0 (30.6°C) 9.1 (42.7°C) 13.9 (52.5°C) 3.0 (64.0°C)

[a]References: W. K. Lewis and E. D. Ries, *Ind. Eng. Chem.*, **19**, 830 (1927); O. A. Uyehara and K. M. Watson, *Ind. Eng. Chem.*, **35**, 541 (1943).
[b]Reference: J. L. Tschernitz, S. Bornstein, R. B. Beckmann, and O. A. Hougen, *Trans. Amer. Inst. Chem. Eng.*, **42**, 883 (1946).
[c]Reference: C. Potter and S. Baron, *Chem. Eng. Progr.*, **47**, 473 (1951).

From S. W. Weller, *Amer. Inst. Chem. Eng.*, **2**, 59 (1956).

a quite adequate representation of the rate data, although they are not (nor do they propose to be) strongly based on any fundamental consideration of nonideal surfaces. In fact, the adequacy of both power-law and LH equations can be explained, in part, by the fact that the LH form,

$$r = \frac{aP}{1 + bP}$$

can be approximated by the power-law form,

$$r = cP^n$$

In addition to the skeptic's view of LH correlations, however, there must be additional reasons why this model has been used successfully for the correlation of kinetics in so many different types of catalytic reactions. Two factors seem to be most important. First is the point illustrated in the previous section and cited earlier here as reason for development of power-law forms: the expressions are insensitive to the precise sort of kinetic scheme involved, so many widely differing types of reaction sequences can be shown to give approximate LH forms. The ultimate example of this has been illustrated by Buzzi Ferraris, et al. for the ammonia synthesis reaction [G. Buzzi Ferraris, G. Donati, F. Rejna and S. Carrà, *Chem. Eng. Sci.*, **29**, 1621 (1974)]. Second is the fact that for quite reasonable assumptions concerning the nature of typical nonuniform surfaces, it can be shown that they tend to look in overall behavior like uniform surfaces. This second point has been discussed in detail in Boudart's text, where an example is given in which the heat of formation of surface complexes is a linear function of surface coverage. For a two-step sequence corresponding to (XXI):

$$\ldots R_1 \rightleftharpoons \ldots R_2$$
$$\ldots R_2 \rightleftharpoons \ldots R_1 \qquad\qquad \text{(XXIa)}$$

the normal result corresponding to an ideal surface is

$$r = \frac{n_s(r_1 r_2 - r_{-1} r_{-2})}{r_1 + r_{-1} + r_2 + r_{-2}} \qquad\qquad \text{(3-52)}$$

whereas the result obtained for a surface with linear variation of heats of chemisorption with coverage is

$$r = \frac{\alpha n_s(r_1 r_2 - r_{-1} r_{-2})}{(r_1 + r_{-1} + r_2 + r_{-2})^\beta} \qquad\qquad \text{(3-53)}$$

The slight difference in these two forms due to the exponent in the denominator of equation (3-53) would in practice not strongly affect the results of an overall correlation of rates, so the comparison suggests that ignoring the nonideality of the surface still leads to rate forms that are qualitatively correct.

3.4 Interpretation of the Kinetics of Reactions on Surfaces

While we have devoted considerable effort in chapter 1 to discussion of means for the interpretation of kinetic data, surface reactions are sufficiently different (and sufficiently more complex) from homogeneous reactions to warrant some separate treatment here. The application of statistical techniques to interpretation of catalytic rate data has been the subject of considerable research in recent years; however, again we must limit our considerations to more basic approaches. For some introduction to the statistical methods in application to reactions on surfaces, see the papers by Kittrell, Mezaki, Hunter, and Watson [J. R. Kittrell, R. Mezaki, and C. C. Watson, *Ind. Eng. Chem.*, **57** (12), 18 (1965); *Brit. Chem. Eng.*, **11**, 15 (1966); W. G. Hunter and R. Mezaki, *Amer. Inst. Chem. Eng. J.*, **10**, 315 (1964); J. R. Kittrell, W. G. Hunter, and C. C. Watson, *Amer. Inst. Chem. Eng. J.*, **11**, 105 (1965); **12**, 5 (1966)].

The fundamental problems in interpretation of heterogeneous rate data are the same as for homogeneous reactions: selecting the rate form which best correlates the measured information and determining the values of the associated constants; much of the general philosophy remains the same. Special difficulties arise because of the more cumbersome form of potential rate equations for surface reactions (assuming now that we are dealing with general LH or related forms and not power-law forms), in evaluation of the relative degree of correlation provided by rival models (discrimination), and in evaluation of the multiple parameters that may be involved from a limited amount of experimental information (estimation). The design of experiments to measure the kinetics of reactions on surfaces is, as a result, much more of an individual affair, differing from one system to another and depending in large measure on what the investigator knows or suspects about the nature of the reaction from other sources or from preliminary experiments. By way of contrast to interpretation of homogeneous kinetics, there are, for example, no conveniently applicable general procedures for testing for reaction order or for the use of reaction equilibrium information in determination of kinetic constants.

From the typical forms illustrated in Table 3.1 it can be seen that a number of different relationships are indicated between the partial pressure of various reactants and products and the rate of reaction. As a result, pressure or concentration is an important variable in experimentation on the kinetics of surface reactions. Conversely, the constants appearing in these equations are all strong functions of temperature, and their appearance both in products and ratios makes interpretation of the temperature dependence of rate (in terms of the individual constants) a sometimes troublesome problem. Related to the use of pressure or concentration as an experimental variable is the use of conversion level as a means for separately determining reactant and product adsorption effects on rate. In particular, the use of initial rate experiments, as discussed in chapter 1, is a convenient method to use when starting a study of the kinetics of a surface reaction.

To illustrate the use of initial rate data, consider the bimolecular reaction (XX), $A + B \rightleftharpoons C + D$. The rate equation for surface reaction control is

$$-r_A = \frac{k'_{S1} K_A K_B P_A P_B - k'_{S2} K_C K_D P_C P_D}{(1 + K_A P_A + K_B P_B + K_C P_C + K_D P_D)^2} \qquad (3\text{-}30)$$

Under initial rate conditions there are no products present and $P_C = P_D \approx 0$. Equation (3-30) becomes

$$(-r_A)_0 = \frac{k'_{S1} K_A K_B P_A P_B}{(1 + K_A P_A + K_B P_B)^2}$$

where $(-r_A)_0$ is the initial rate of reaction of A.

Now let us consider a particular type of initial rate experiment, one in which $P_A = P_B$, or, in terms of total pressure, $P_A = P_B = \frac{1}{2} P_T$. The initial rate becomes

$$(-r_A)_0 = \frac{k'_{S1} K_A K_B P_T^2}{4[1 + 0.5(K_A + K_B) P_T]^2} \qquad (3\text{-}54)$$

Writing equation (3-54) in terms of combined constants a and b:

$$(-r_A)_0 = \frac{a P_T^2}{(1 + b P_T)^2} \qquad (3\text{-}54a)$$

where $a = k'_{S1} K_A K_B / 4$ and $b = 0.5(K_A + K_B)$.

Similar derivations using the Hougen and Watson procedure give, for an adsorption of A rate controlling,

$$(-r_A)_0 = \frac{a' P_T}{1 + b' P_T} \qquad (3\text{-}55)$$

or for an Eley–Rideal mechanism in which one of the reactants is not adsorbed [reaction (6), Table 3.1],

$$(-r_A)_0 = \frac{a'' P_T^2}{1 + b'' P_T} \qquad (3\text{-}56)$$

where a', a'', b', and b'' are combined constants analogous to a and b.

There is a quite different dependence of $(-r_A)_0$ on total pressure according to the last three equations above, the general form of which is shown in Figure 3.8. Curves of this type for a number of reaction schemes with various rate-controlling steps have been given by Hougen[1] (O. A. Hougen, *Reaction*

---

[1] Most authors include in such comparisons the initial rate dependence corresponding to a product desorption controlling step. The result gives $(-r_A)_0$ independent of P_T, which seems intuitively correct. However, the interpretation of initial rate data in terms of a product step seems a contradiction, and it is best to study product steps with specific experiments involving variation of conversion level (i.e., product partial pressures). "Them that asks no questions isn't told a lie" (Rudyard Kipling).

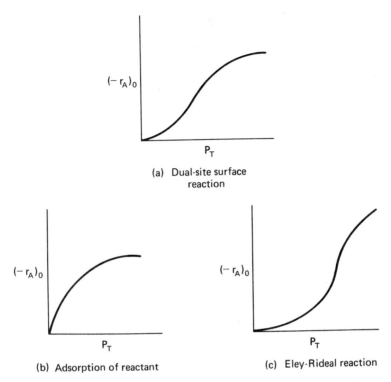

(a) Dual-site surface reaction

(b) Adsorption of reactant

(c) Eley-Rideal reaction

Figure 3.8 Comparison of initial rate dependence on total pressure for various controlling steps in the reaction $A + B \rightleftharpoons C + D$

Kinetics in Chemical Engineering, Chem. Eng. Progr. Monograph Ser., no. 1, 1951); it can be seen that such initial rate experiments are useful for screening purposes in rejection of obviously inappropriate rate forms. Not much more than this can realistically be expected, however, since the difference in shape between various cases is sometimes rather subtle granted the experimental range of P_T it is possible to cover and the precision of the measured rate data. For example, distinguishing between cases (a) and (c) in Figure 3.8 depends on being able to distinguish between a rate proportional to P_T at high values of P_T, and a rate approaching zero order in P_T in this region. Distinguishing between (a) and (b) depends on being able to define experimentally the presence or absence of the characteristic inflection of (a).

From the results of such initial rate experiments, hopefully one will have gained sufficient insight into the reaction system to permit intelligent planning of further experiments from which the numerical values of the constants for rival equations may be determined which give the best fit in each case. Then some sort of overall comparison of the individual equations based on goodness of fit to the experimental data and physical reasonableness can be used for a final selection. There are a number of different means for evaluating constants corresponding to best fit and for comparing relative goodness of fit, generally involving nonlinear methods, as discussed in the papers of Kittrell et al. However, the methods of linear least squares may be used for parameter evaluation and comparative fit, *at least as a first estimate*, as in the following illustration of the procedure for the initial rate form of equation (3-30). We may write this in a form convenient for the evaluation of constants by rearranging:

$$y = \sqrt{\frac{P_A P_B}{(r_A)_0}} = \frac{1 + K_A P_A + K_B P_B}{(k'_{S_1} K_A K_B)^{1/2}} = C_1 + C_2 P_A + C_3 P_B \qquad (3\text{-}57)$$

where $C_1 = (k'_{S_1} K_A K_B)^{-1/2}$, $C_2 = (K_A/k'_{S_1} K_B)^{1/2}$, and $C_3 = (K_B/k'_{S_1} K_A)^{1/2}$. If we have n measurements of y for various conditions of P_A and P_B, equation (3-57) becomes the set

$$y_1 = C_1 + C_2 (P_A)_1 + C_3 (P_B)_1$$
$$y_2 = C_1 + C_2 (P_A)_2 + C_3 (P_B)_2$$
$$\cdot \qquad\qquad\qquad\qquad \cdot$$
$$\cdot \qquad\qquad\qquad\qquad \cdot \qquad\qquad (3\text{-}57a)$$
$$\cdot \qquad\qquad\qquad\qquad \cdot$$
$$y_n = C_1 + C_2 (P_A)_n + C_3 (P_B)_n$$

and by least squares we wish to find the values of C_1, C_2, and C_3 which minimize the sum of squares of residuals, $v_1, v_2, \ldots, v_n$, defined as

$$v_1 = C_1 + C_2 (P_A)_1 + C_3 (P_B)_1 - y_1$$
$$v_2 = C_1 + C_2 (P_A)_2 + C_3 (P_B)_2 - y_2$$
$$\cdot \qquad\qquad\qquad\qquad\qquad \cdot$$
$$\cdot \qquad\qquad\qquad\qquad\qquad \cdot \qquad\qquad (3\text{-}58)$$
$$\cdot \qquad\qquad\qquad\qquad\qquad \cdot$$
$$v_n = C_1 + C_2 (P_A)_n + C_3 (P_B)_n - y_n$$

The condition for minimization is

$$\frac{\partial\left(\sum_{i=1}^{n} v_i^2\right)}{\partial C_1} = \frac{\partial\left(\sum_{i=1}^{n} v_i^2\right)}{\partial C_2} = \frac{\partial\left(\sum_{i=1}^{n} v_i^2\right)}{\partial C_3} = 0 \qquad (3\text{-}59)$$

yielding the following equations, called the *normal equations* for the set of (3-57a):

$$C_1 n + C_2 \sum_{i=1}^{n} (P_A)_i + C_3 \sum_{i=1}^{n} (P_B)_i - \sum_{i=1}^{n} y_i = 0$$

$$C_1 \sum_{i=1}^{n} (P_A)_i + C_2 \sum_{i=1}^{n} (P_A)_i^2 + C_3 \sum_{i=1}^{n} (P_A)_i (P_B)_i - \sum_{i=1}^{n} (P_A)_i y_i = 0 \quad (3\text{-}60)$$

$$C_1 \sum_{i=1}^{n} (P_B)_i + C_2 \sum_{i=1}^{n} (P_A)_i (P_B)_i + C_3 \sum_{i=1}^{n} (P_B)_i^2 - \sum_{i=1}^{n} (P_B)_i y_i = 0$$

In equations (3-60) the only unknowns are C_1, C_2, and C_3, which can be evaluated by any convenient method for solving simultaneous linear algebraic equations. Individual values for k'_{S1}, K_A, and K_B can then be computed. An obvious method for comparing the fit afforded a given set of data by various rate forms is then to use the least-squares constants so determined to calculate the sum of squares of residuals corresponding to the different forms and to compare their magnitudes.

For a given rate expression the choice of a form of equation for least-squares evaluation, such as equation (3-57), is largely a matter of convenience. For example, we could have chosen $y = \sqrt{1/(r_A)_0}$, which would give an entirely different form of expression for sum of squares of residual minimization. *In general, the residuals generated by different forms of the same rate equation will differ from each other and hence also the least-squares constants.* This means in the present illustration, for example, that the values of k'_{S1}, K_A, and K_B evaluated according to $y = \sqrt{P_A P_B/(r_A)_0}$ and $y = \sqrt{1/(r_A)_0}$, respectively, will not be the same. Also, since we have not worked with the residuals of rate directly but with some function of the rate, the least-squares constants so determined do not in general correspond to the best fit of the rate itself. In spite of such reservations and drawbacks associated with the linear least-squares procedure, however, it remains a convenient method for at least first estimates for comparison of rate equations and evaluation of parameters. The utility of the method is reinforced by the fact that most modern computation facilities offer system programs for linear least-squares analysis of experimental data, so the apparently tedious calculations involved in equations (3-60) are, in fact, effortless.

There have been many suggestions made concerning methods by which the meaningfulness of the constants determined in an LH interpretation of rates can be evaluated. Certainly, negative values for any of the constants would cast considerable doubt on the validity of the rate equation, although in such an event one must be sure that the negative value is not an artifact created by a fit to experimental data with considerable scatter. This particularly can happen in the evaluation of adsorption constants contained in the denominator of the rate equation when the species in question is very weakly adsorbed and the true value of the constant is essentially zero.

A very useful method for determining the reasonableness of constants estimated by LH analysis has been advanced by Boudart, Mears, and Vannice [M. Boudart, D. E. Mears, and M. A. Vannice, *Ind. Chim. Belge*, **32**, 281 (1967)]. The method is based on the compensation effect, often noted in the kinetics of catalytic reactions on a series of related catalysts, in which there is observed a linear relationship between the logarithm of the preexponential factor of the rate equation and the activation energy. The effect was first noted by Constable [F. H. Constable, *Proc. Roy. Soc. (London)*, **A108**, 355 (1925)] for ethanol dehydrogenation on a series of copper catalysts reduced at different temperatures; this correlation is shown in Figure 3.9. Compensation effects have been reported for the rates of a large number of catalytic reactions and, more important for our present purposes, for physical adsorption equilibria by Everett [D. H. Everett, *Trans. Faraday Soc.*, **46**, 942 (1950)] in terms of a linear relationship between the entropy and enthalpy changes on adsorption. Thus,

$$(\Delta S_a)^\circ = m(\Delta H_a)^\circ + b \tag{3-61}$$

where $(\Delta S_a)^\circ$ is the entropy change on adsorption, most conveniently referred

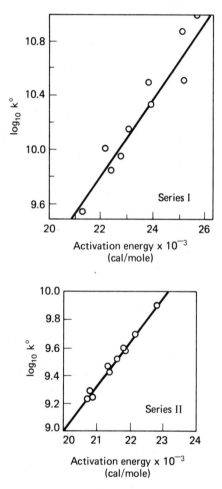

Figure 3.9 Example of the compensation effect: data of Constable for ethanol dehydration on a series of copper catalysts reduced at different temperatures (after F. H. Constable, *Proc. Roy. Soc. (London)*, **A108**, 355 (1925); with permission of The Royal Society)

to a standard state of 1 atm at the temperature of adsorption, and $(\Delta H_a)^\circ$ the corresponding enthalpy change. From experimental data on the physical adsorption of a large number of gases at less than monolayer coverage on charcoal, Everett obtained the correlation

$$(\Delta S_a)^\circ = 0.0014(\Delta H_a)^\circ - 12.2 \tag{3-62}$$

for $(\Delta S_a)^\circ$ in cal/g mole-°K and $(\Delta H_a)^\circ$ in cal/g mole. Following these ideas, Boudart et al. reported the results of a $(\Delta S_a)^\circ - (\Delta H_a)^\circ$ analysis based on a large number of chemisorbed systems, using adsorption constant data reported from LH interpretations in the literature, which were deemed to be reasonable results. Writing these equilibrium constants in thermodynamic terminology,

$$K = e^{(\Delta S_a)^\circ/R}e^{-(\Delta H_a)^\circ/RT} \tag{2-107a}$$

we have an expression that permits calculation of $(\Delta S_a)^\circ$ from experimental information on K and values of $-(\Delta H_a)^\circ$, which may be determined for the particular chemisorption involved. The results of this analysis, in comparison with Everett's correlation, are shown in Figure 3.10. While it would appear that the chemisorption data are not linearly correlated, it seems reasonable to establish the Everett correlation as a limiting case for the $(\Delta S_a)^\circ - (\Delta H_a)^\circ$ relationship, since almost all the data fall below the line. The result of Figure 3.10 thus provides some guide as to reasonable values of $(\Delta S_a)^\circ$ in chemisorbed-reacting systems and can be used as a further test of the meaningfulness

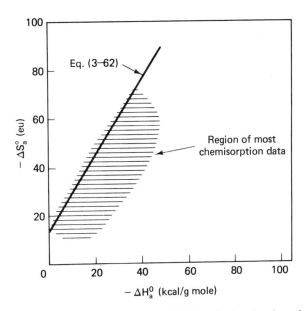

Figure 3.10 Correlation of ΔS_a° and ΔH_a° results for chemisorption

of the adsorption constants appearing in LH correlations of reaction rates. We can, in fact, establish some rules concerning reasonable values for $(\Delta S_a)°$. First, since the chemisorbed layer is a more ordered state than the gas phase, $(\Delta S_a)°$ should be negative. Second, $(\Delta S_a)°$ should be smaller than the standard-state entropy of the gas phase, $(S_g)°$, since the entropy change on adsorption cannot be larger than the entropy of the nonadsorbed state. These two rules are rather obvious ones, which Boudart et al. term "strong rules" for the reasonableness of $(\Delta S_a)°$ determined from kinetic interpretation. A "weak rule," based on the calculation

$$(\Delta S_a)° = R \, \ell\mathrm{n} \left(\frac{V_g}{V_c}\right) \tag{3-63}$$

where V_g is the molar volume of gas at 1 atm and adsorption temperature and V_c the critical volume of the liquid (adsorbed state) states that $(\Delta S_a)°$ should be of the order of -10 eu.

3.5 Decline of Surface Activity: Catalyst Deactivation

One particular aspect of reactions on surfaces that is not encountered so often in homogeneous reactions is a progressive decrease in the activity of the surface with its time of utilization. The reasons for this are numerous, but we will divide them into three general categories:

1. *Poisoning:* loss of activity due to strong chemisorption of a chemical impurity on the active sites of the surface, denying their access by reactant molecules. This should not be confused with inhibition as expressed by terms in the denominator of LH rate expressions.
2. *Coking or fouling:* loss of activity, normally in reactions involving hydrocarbons, due to reactant or product degradation producing a carbonaceous residue on the surface which is inactive for catalysis.
3. *Sintering:* loss of activity due to a decrease in active surface per unit volume of catalyst, normally the result of excessively high temperatures.

A vast amount of effort has been expended over the years in investigation of these types of deactivation as they are encountered in catalytic reactions and catalysts of technological importance. The neophyte chemical engineer is often amazed at the fact that many reaction-system process designs are dictated by the existence of catalyst deactivation, as are process operation and optimization strategies. The automobile exhaust converter is one recent, important example of catalyst/reaction design largely determined by the existence of deactivation phenomena.

In Table 3.3 are given some examples of the three major categories of deactivation. The literature cited generally refers to earlier work which has turned out to be definitive of the classifications used. More detail is available in a brief review [J. B. Butt, *Advan. Chem.*, **109**, 259 (1972)]. Our interests here will be confined to reaction networks which may reasonably be used to represent the various mechanisms of deactivation, and to plausible means for correlating the kinetics of deactivation.

TABLE 3.3
SOME EXAMPLES OF CATALYST DEACTIVATION

Reference[a]	System	Activity/Time	Comments
1	Poisoning of Pt by metals, S, N, hydrogenation reactions	Linear, exponential	First systematic investigation of impurity poisoning effects on noble metals
2	Poisoning of SiO_2 and SiO_2/Al_2O_3 by organic bases, cracking reactions	Exponential	Demonstrated acidic nature of these oxides, developed activity correlation based on poisoning behavior
3	Poisoning of different types of Al_2O_3 with various alkali metals, isomerization and dehydration reactions	Linear, exponential	Showed specificity of poisoning behavior to both chemical nature of poison and reaction
4	Coking of metal-oxide cracking catalysts	Power law, time on stream	Developed correlation for coke on catalyst versus time of reaction
5	Coking of SiO_2/Al_2O_3 and zeolite cracking catalysts	Exponential, time on stream	Deactivation and kinetic models for catalytic cracking
6	Sintering in reforming on Pt/Al_2O_3	Hyperbolic	Preferential deactivation of one function of a bifunctional catalyst
7	Sintering, primarily supported Pt	Many	Review paper

[a] 1, E. B. Maxted, *Advan. Catalysis*, **3**, 129 (1951); 2, G. A. Mills, E. R. Boedeker, and A. G. Oblad, *J. Amer. Chem. Soc.*, **72**, 1554 (1950); 3, H. Pines and W. O. Haag, *J. Amer. Chem. Soc.*, **82**, 2471 (1960); 4, A. Voorhies, Jr., *Ind. Eng. Chem.*, **37**, 318 (1945); 5, V. W. Weekman, Jr., *Ind. Eng. Chem. Proc. Design Devel.*, **7**, 90 (1968); **8**, 388 (1969); V. W. Weekman, Jr., and D. M. Nace, *Amer. Inst. Chem. Eng. J.*, **16**, 397 (1970); D. M. Nace, S. E. Voltz, and V. W. Weekman, Jr., *Ind. Eng. Chem. Proc. Design Devel.*, **10**, 530, 538 (1971); 6, H. J. Maat and L. Moscou, *Proc. Int. Congr. Catalysis, 3rd*, p. 1277, North-Holland, Amsterdam, 1965; 7, S. E. Wanke and P. C. Flynn, *Cat. Rev.—Sci. Engr.*, **12**, 93 (1975).

Much of the experimental work developing reaction networks involved in deactivation has successfully employed schemes very similar to our Type

I and Type III "nearly complex" reactions. If we let S be an active site on the surface, independent chemical poisoning in the simplest example can be represented by

$$A + S \longrightarrow B + S$$
$$L + S \longrightarrow L \cdot S \tag{XXII}$$

where in this case L is the poison that removes active sites from the surface via irreversible chemisorption. It can be seen that this is very similar to the Type I scheme. Networks describing deactivation via coking may be more complex than this, since in many hydrocarbon reactions both reactants and products are capable of forming the carbonaceous deposits. A plausible, simple example of this is

$$A + S \longrightarrow B + S$$
$$A + S \longrightarrow A \cdot S \tag{XXIII}$$
$$B + S \longrightarrow B \cdot S$$

where A · S and B · S represent the deactivated surface. This is reminiscent of the Type III scheme, with A the reactant, B the intermediate desired product, and A · S and B · S corresponding to C of Type III. Sintering does not fall so easily in this category, since it is normally a thermally activated process and generally independent of reactants or products; another way to think of it is that sintering is a physical phenomenon and so is not properly described by chemical reaction networks.

There is an important variation on these mechanisms of deactivation when one is dealing with bifunctional catalysts, that is, surfaces that possess more than one type of catalytic function. (An example of this is given in problem 13 for n-pentane isomerization on Pt/Al_2O_3.) Here one might envision one reaction being carried out on, say, the X function of the catalyst, and a second on the Y function, and the reactions may be either independent (parallel) or sequential. Again there is resemblance to Types I and III, but the representation is more complex. For the parallel sequence:

Main:
$$A + S_1 \xrightarrow{X} B + S_1$$
$$C + S_2 \xrightarrow{Y} D + S_2$$

Poisoning:
$$L + S_1 \xrightarrow{X} L \cdot S_1$$
$$M + S_2 \xrightarrow{Y} M \cdot S_2 \tag{XXIV}$$

and for the series sequence:

Main:
$$A + S_1 \xrightarrow{X} B + S_1$$
$$B + S_2 \xrightarrow{Y} C + S_2$$

Poisoning:
$$L + S_1 \xrightarrow{X} L \cdot S_1$$
$$M + S_2 \xrightarrow{Y} M \cdot S_2 \tag{XXV}$$

The basic importance of these schemes is that they indicate clearly how deactivation may seriously interfere with the selectivity, as well as the activity, of the surface. Consider (XXV): if the Y function is poisoned to a greater extent than the X function, there will obviously come a time when the product C is no longer observed. If the converse holds true, the reaction will eventually shut down completely, even though Y activity remains, since there is no B intermediate. In supported metal catalysts the metal is often preferentially deactivated by sintering over a long period and there is a gradual change in product selectivity; such a situation is illustrated by the work of Maat and Moscou in Table 3.3, and is an example of a process operation ultimately being dictated by deactivation. Here, when the product selectivity for dehydrocyclization drops below a certain level, operation will have to be discontinued and a fresh charge of catalyst introduced.

Some approaches to the modeling of the kinetics of catalyst deactivation are also suggested by the results of Table 3.3. We see there observations of the variation of catalyst activity with time which in some cases are linear, exponential, or hyperbolic. Hopefully, we remember from chapter 1 that these types of temporal variations are the fingerprints of zero-, first-, and second-order reactions, respectively, so the suggestion is that catalyst deactivation kinetics may be represented by simple power-law forms. Now, how might this variation be incorporated into the rate law for a surface reaction? Let us reconsider the isomerization example of scheme (XVIII) on an ideal surface but with the surface reaction step slow and rate-controlling:

$$A + S \rightleftharpoons A \cdot S$$
$$A \cdot S \longrightarrow B \cdot S \qquad \text{(XVIII)}$$
$$B \cdot S \rightleftharpoons B + S$$

The surface sites represented by S we will consider to be at an activity level s lower than that of the undeactivated surface, where $s = 1$ would represent fresh surface and $s = 0$ completely deactivated surface. In this case the kinetics of the three steps of XVIII are

$$k_{a_A} P_A (1 - \theta_A - \theta_B)s = k_{d_A} \theta_A s \qquad (3\text{-}64)$$

$$(-r_s) = k_s' \theta_A s \qquad (3\text{-}65)$$

$$k_{d_B} \theta_B s = k_{a_B} P_B (1 - \theta_A - \theta_B)s \qquad (3\text{-}66)$$

As shown before, the development of a rate equation here depends on the development of an expression for θ_A. If we follow the procedures detailed for equations (3-6) to (3-10), it is clear that the net adsorption and desorption rate constants, such as $k_{a_A}s$ and $k_{d_A}s$, will always appear in ratio, and the activity variable s will divide out. The resulting overall rate equation then becomes

$$(-r_A) = \frac{k'_s s K_A P_A}{1 + K_A P_A + K_B P_B} \tag{3-67}$$

The effect of deactivation on the kinetics of the main reaction is thus modeled by a single activity factor, s, multiplying the intrinsic surface reaction rate constant. The rate of deactivation, as suggested by the data of Table 3.3, would be expressed in a separate rate equation, the precise form depending on the mechanism of deactivation. For example, if poisoning were the mechanism of deactivation in a scheme such as (XXII), then

$$\frac{-ds}{dt} = (-r_s) = k_{da} s P_L \tag{3-68}$$

This approach is termed *separable deactivation*, since the activity factor s is carried along as separable in the formulation of equation (3-67), and was originally proposed by Szepé and Levenspiel (S. Szepé and O. Levenspiel, *Proc. European Fed., 4th Chem. Reaction Eng., Brussels*, Pergamon Press, London, 1970).

An alternative approach to the correlation of catalyst deactivation kinetics has been widely employed for coking mechanisms. Here the precise reactions that cause deactivation may be more complex or more difficult to identify than in chemical poisoning, and one may not write such precise expressions as equation (3-68) for $(-r_s)$. The practice has been generally to relate activity to the time of utilization or "time on stream," and a number of mathematical forms have been employed. The best known is the *Voorhies correlation* [A. Voorhies, Jr., *Ind. Eng. Chem.*, **37**, 318 (1945)], which relates the weight of coke on catalyst to a power of time on stream:

$$C_c = A t_c^n \tag{3-69}$$

where A and n are emperically determined constants. Other forms that have been used are exponential and hyperbolic. It is interesting to note that these correlations generally refer to the amount of coke, C_c, not the activity factor s, so that normally it is necessary to develop additional information of the form

$$s = f(C_c) \tag{3-70}$$

Again, various relationships have been reported for equation (3-70), including linear, hyperbolic, and exponential functions.

As a final comment on catalyst deactivation, we add the following. The existence of catalyst deactivation means that there is a change in the overall reaction rate with time; this has the simple but profound effect of changing an entire area of chemical kinetics from a steady-state problem (the ideal world as set forth in Table 3.1) into an unsteady-state problem [the real world of equations (3-67) and (3-68)]. Ponder the point for now; its impact will be explored further in chapter 7.

EXERCISES FOR CHAPTER 3

Section 3.1

1. Calculate the rate at which nitrogen molecules at 1 atm and 100°C strike an exposed surface.

2. Compare the rate computed in problem 1 with the rate of chemisorption of nitrogen molecules, at the same conditions of temperature and pressure, on an hypothetical surface with site density, n_S, of 10^{15} cm^{-2} at half-monolayer surface coverage. The activation energy for chemisorption is 5 kcal/mole, and $\sigma = 0.1$.

3. Using a TST approach and assuming the adsorbed molecule to be completely localized (only vibrational degrees of freedom), show that σ in equation (3-3) is proportional to the ratio of the vibrational partition function of the adsorbate to the vibration plus rotation partion functions of the nonadsorbed gas molecule.

4. Calculate the equilibrium surface coverage according to the Langmuir isotherm for the nitrogen example of problems 1 and 2. The heat of adsorption is 10 kcal/mole, the sticking probability, σ, is unity, and $k_a^\circ \approx 10^{30}$ sec^{-1}.

5. Derive the Langmuir isotherm expressions for the equilibrium surface coverages, θ_A and θ_B, of two gases A and B competitively adsorbed on the same surface [equations (3-13)].

6. The following data were reported by Taylor and Williamson [H. S. Taylor and A. T. Williamson, *J. Amer. Chem. Soc.*, **53**, 2168 (1931)] for the chemisorption of hydrogen on a mixed manganous/chromic oxide powder at 305 and 444°C.

$T = 305°C$		$T = 444°C$	
Pressure (mm Hg)	Vol. ads. (cm³)	Pressure (mm Hg)	Vol. ads. (cm³)
44	156.9	3	57.1
51	160.8	22	83.3
63	163.6	48	95.0
121	167.0	77	98.1
151	169.6	165	100.9
230	171.1		
269	171.6		

Test these data for adherence to the Langmuir isotherm. What is the monolayer volume in each case? Are these consistent results?

7. Using a procedure analogous to that employed to derive the Freundlich isotherm, obtain the Temkin isotherm equation, equation (3-25).

8. Below are the data of Emmett and Brunauer [P. H. Emmett and S. Brunauer, *J. Amer. Chem. Soc.*, **56**, 35 (1934)] for the adsorption of nitrogen on powdered iron catalyst at 396°C.

Pressure (mm Hg)	Vol. ads. (cm³) (STP)
25	2.83
53	3.22
150	3.69
397	4.14
768	4.55

From these data can you distinguish which of the three isotherms—Langmuir, Freundlich, or Temkin—is best obeyed?

Section 3.2

9. Following the procedure described in section 3.2, derive the rate equations for reactions (2) and (6) in Table 3.1.
10. Derive a rate equation for reaction scheme (XX) if the rate of adsorption of reactant A is rate-controlling. Repeat for the rate of desorption of C rate-controlling.
11. Using the general equation for rate, equation (3-45), derive equation (3-46) for the rate of the isomerization reaction (XVIII).

Section 3.3

12. Using the general pssh approach and the restriction equation (3-40), derive the rate equation for reaction (2) in Table 3.1.

Section 3.4

13. Sinfelt, Hurwitz, and Rohrer [J. H. Sinfelt, H. Hurwitz, and J. C. Rohrer, *J. Phys. Chem.*, **64**, 892 (1960)] studied the kinetics of the isomerization of *n*-pentane over a platinum-on-alumina catalyst in the presence of excess hydrogen. Overall:

$$H_2 + n\text{-}C_5 \xrightarrow[\text{Al}_2\text{O}_3]{\text{Pt}} H_2 + i\text{-}C_5$$

They postulate a mechanism whereby the *n*-C$_5$ is chemisorbed on the Pt, simultaneously being dehydrogenated to the *n*-C$_5$ olefin. The chemisorbed *n*-pentene species then desorbs from the Pt sites and is readsorbed on the acidic sites of the alumina, where isomerization to *i*-pentene occurs. The *i*-pentene is then desorbed and readsorbed on the Pt site, where it is rehydrogenated to *i*-pentane and desorbed.

Assuming the isomerization is rate-controlling and irreversible, write the reaction scheme representing the mechanism above. Derive the equation for the rate of *n*-pentane isomerization if all species involved obey Langmuir adsorption isotherms.

14. The disproportionation of propylene on supported tungsten oxide catalysts is thought to proceed via a cyclobutane intermediate as follows:

$$2C_3^=(g) \rightleftharpoons \begin{array}{c} C-C-C \\ | \quad | \\ C-C-C(ads) \end{array}$$

$$\begin{array}{c} C-C-C \\ | \quad | \\ C-C-C(ads) \end{array} \rightleftharpoons C_4^=(ads) + C_2^=(ads)$$

$$C_4^=(ads) + C_2^=(ads) \rightleftharpoons C_4^=(g) + C_2^=(g)$$

In this view the decomposition of the cyclobutane intermediate would be the rate-determining step. Alternative views hold that the first step above should be broken into two steps, adsorption of $C_3^=$ and surface reaction of $2C_3^=(ads)$ to form the intermediate, or that the formation of intermediate is via reaction of $C_3^=(g)$ with $C_3^=(ads)$. In both these cases the decomposition of intermediate is rapid.

(a) Develop rate equations for these three views of the reaction mechanism.

(b) Simplify your results for the initial rate of reaction, (i.e., $P_{C_4^=}$ and $P_{C_2^=}$ are small).

(c) Examine the three rate equations in view of the following results [U. Hattikudur and G. Thodos, *Advan. Chem.*, **133**, 80 (1974)]:

$T\,(°F)$	$P_{C_3^=}\,(atm)$	Initial rate $(g\ moles/g\ cat\text{-}hr)$
650, 750, 850	2.17, 2.09, 2.06	0.58, 2.40, 6.80
	6.98, 2.52, 9.41	2.30, 2.48, 11.39
	9.67, 6.89, 17.58	2.98, 5.30, 12.29
	35.86, 17.82, 34.37	5.13, 7.46, 12.92
	50.85, 34.86, 48.91	5.42, 8.14, 13.37
	65.24, 49.55, —	5.86, 8.60, —

15. In Figure 3.11 are given some experimental results of Kehoe and Butt (J. P. G. Kehoe and J. B. Butt, *J. Appl. Chem. Biotechnol.*, **1972**, 23) on the initial rates of benzene hydrogenation over a nickel–kieselguhr catalyst in an excess of hydrogen at low temperatures. Determine a consistent form of Langmuir–Hinshelwood expression to correlate these data and obtain the values of the associated rate and equilibrium constants and their temperature dependence.

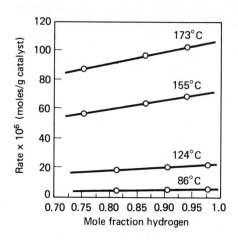

(a) Rate of Hydrogenation as a Function
of Hydrogen Concentration: P = 760 Torr

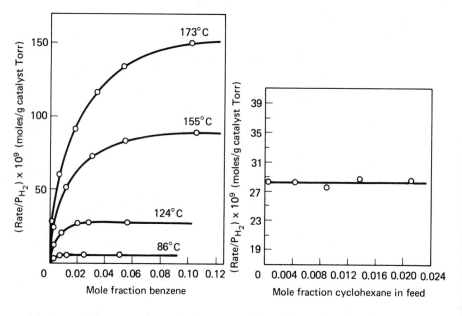

(b) Rate of Hydrogenation as a Function
of Benzene Concentration: P = 760 Torr

(c) Product Inhibition Effects at Low
Conversions: P = 760 Torr, T = 124°C

Figure 3.11 Initial rate data on benzene hydrogenation over Ni–kieselguhr
(after J. P. G. Kehoe and J. B. Butt, *J. Appl. Chem. Biotechnol.*, **1972**, 23;
with permission of the Society of Chemical Industry)

16. The mechanism proposed by Topchieva [K. V. Topchieva, K. Yun-Pun and I. V. Smirnova, *Advan. Catalysis*, **9**, 799 (1957)] for ethanol dehydration over alumina is

$$
\begin{array}{cc}
& \text{H} \\
& \text{O} \qquad \text{C}_2\text{H}_5\text{O} \\
& | \qquad\qquad | \\
\text{(1)} \quad \text{C}_2\text{H}_5\text{OH} + & \text{—Al—} \rightleftharpoons \text{—Al—} + \text{H}_2\text{O}
\end{array}
$$

$$
\begin{array}{cc}
& \text{H} \\
\text{C}_2\text{H}_5\text{O} & \text{O} \\
| & | \\
\text{(2)} \qquad \text{—Al—} \rightleftharpoons \text{—Al—} + \text{C}_2\text{H}_4 \\
| \\
\text{H} \\
\text{O} \\
\end{array}
$$

The designated active site, —Al—, is consumed in the first step and regenerated in the second, thus forming a closed sequence.

Derive an equation for the rate of reaction of alcohol assuming that:

(a) Both reactions occur at the same rate.

(b) Reaction (1) is at equilibrium and (2) is rate-controlling and irreversible.

17. The following initial rates in (lb moles A reacted/hr-lb catalyst) have been reported for the reaction

$$A \longrightarrow C + D$$

	$P_A(atm)$										
$T(°F)$	0.5	1.0	2.0	2.5	3.0	4.0	5.0	7.0	8.0	9.0	13.0
550	0.008	0.012	0.015	—	0.017	0.017	—	0.014	—	0.013	0.011
575	0.016	0.024	—	—	0.028	—	0.026	0.022	—	0.019	—
600	0.041	0.042	—	—	0.043	0.042	0.042	0.042	—	0.036	0.027
650	0.113	—	—	—	0.114	—	—	0.116	—	0.109	0.098
700	0.192	0.195	0.190	0.189	0.193	0.188	0.183	—	0.195	0.197	0.190

Determine a consistent explanation for these data, giving the rate equations that are applicable and evaluating all pertinent constants. The desorption rate of D is slower than that of C under all conditions reported here.

18. The dehydrogenation of methylcyclohexane to toluene was studied by Sinfelt et al. [J. H. Sinfelt, H. Hurwitz, and R. A. Shulman, *J. Phys. Chem.*, **64**, 1559 (1960)] over a 0.3% Pt on Al_2O_3 catalyst. The temperature range from 315 to 372°C was investigated, with methylcyclohexane partial pressures varying from 0.07 to 2.2 atm, and hydrogen partial pressures from 1.1 to 4.1 atm. Experimental rate data were well correlated by an equation of the form

$$r = \frac{k'bP_M}{1 + bP_M}$$

where r is the rate of the dehydrogenation reaction and P_M the pressure of methylcyclohexane. The following values were obtained for the constants:

$T\,(°C)$	k'	$b\,(atm^{-1})$
315	0.013	27
344	0.043	11
372	0.154	3

There are difficulties in associating the rate correlation with a Langmuir–Hinshelwood mechanism, however, since the temperature dependence of the adsorption constant b corresponds to a heat of adsorption of about 30 kcal/mole, far too high for this system in light of independent adsorption information. Further, the absence of any product adsorption term in the denominator is suspicious, since aromatics are known to be more strongly adsorbed on the catalyst then are saturated compounds.

Show that a similar form of equation may be derived assuming the reaction scheme to be

$$M \xrightarrow{k_1} \underset{\substack{\text{dehydrogenation}\\\text{steps}}}{[M_1]} \longrightarrow T(ads) \xrightarrow{k_2} T(g)$$

where the dehydrogenation steps are very rapid and the adsorbed toluene intermediate is at steady state. The rate of desorption of toluene is controlling. Show also that the temperature dependence of the constants of the scheme above also provide a much more reasonable interpretation of the experimental results.

19. Synthesis gas, primarily methane and hydrogen, produced by the steam reforming of hydrocarbons still contains about 0.5 mole % CO_2 after scrubbing to remove CO and CO_2. It is important to remove this impurity before using the gas in subsequent processes (i.e., ammonia synthesis), and this is normally accomplished by hydrogeneration to methane on a Ni/Al_2O_3 catalyst:

$$CO_2 + 4H_2 \xrightarrow{Ni} CH_4 + H_2O$$

Figure 3.12 gives some data for the rate of methanation of CO_2. Hydrogen is present in great excess in these experiments.

each other, in that definition of a state of macromixing does not, for example, define a corresponding level of micromixing.

Well-defined limits of macro- and micromixing can be obtained in a number of instances, and these serve to define corresponding ideal reactor types. Deviations of mixing from these limits are sometimes termed *nonideal flows*. Since it is difficult to define a measure for quantities such as the degree of micromixing or, indeed, to make measurements on the hydrodynamic state of the internals of a reactor system, extensive use has been made of models that describe the observable behavior in terms of external measurements. As in any other kind of modeling, these models may not be correct in detail and a number of different forms, each possessing different degrees of reality with respect to the phenomena being modeled, may equally well be used over a sufficiently small range of operating conditions. Mixing models, both for ideal and nonideal flows, lead directly to chemical reactor models; however, before treating these let us examine in a more general way the concepts of micro- and macromixing.

4.1   Reactions in Mixed or Segregated Systems

In order to visualize physically what is meant by molecular level or micromixing, consider a "point"-volume element of fluid within the reactor which is just large enough to possess the average value of an intensive property such as concentration at the particular location considered [P. V. Danckwerts, *Chem. Eng. Sci.*, **8**, 93 (1958)]. All molecules within a point-volume element have zero age when entering the reactor, and all have zero time to remain within the reactor (life expectancy) when the element exits. *The degree of micromixing defines what happens to the individual molecules of a point-volume element in the time between its entrance and exit from the reactor.*

We can readily envision two extremes. First, all the molecules within the point element on its entrance to the reactor remain within it until exit, as if the element were surrounded by a molecularly impermeable barrier. This is termed *segregated* flow. Second, complete mixing on the molecular level occurs immediately on entrance, which is termed a state of *maximum mixedness* [T. N. Zweitering, *Chem. Eng. Sci.*, **11**, 1 (1959)]. We may define segregated flow more precisely by saying that the variance of the age distribution of molecules within the point element is zero. Detailed analysis of intermediate degrees of micromixing requires the use of various models, a number of which we shall discuss later.

What is the importance of the degree of micromixing to chemical reaction systems? This question is perhaps best answered using a famous example of Danckwerts from the paper cited above. Consider the isothermal, second-order irreversible reaction $A + B \longrightarrow C$, carried out homogeneously and at constant volume. The initial concentrations of A and B, C_{A_0} and C_{B_0}, respectively, are set in the ratio

$$C_{B_0} = C_{A_0}(1 + x) \tag{4-1}$$

Now, if the system is in the state of maximum mixedness, after a fraction f of A is reacted the concentrations are

$$C_A = C_{A_0}(1 - f)$$
$$C_B = C_{B_0}(1 + x - f) \tag{4-2}$$

At the opposite end of the spectrum of micromixing, let us consider the same system, but now consisting of two segregated portions having differing degrees of conversion of A, f_1, and f_2, with the volumes of the two portions in the ratio α to $(1 - \alpha)$. The concentrations of A in the two portions are, obviously,

$$C_{A_1} = C_{A_0}(1 - f_1)$$
$$C_{A_2} = C_{A_0}(1 - f_2) \tag{4-3}$$

and for B:

$$C_{B_1} = C_{A_0}(1 - f_1 + x)$$
$$C_{B_2} = C_{A_0}(1 - f_2 + x) \tag{4-4}$$

The rate of reaction, averaged over the two portions, is

$$r_1 = k\alpha C_{A_0}^2(1 - f_1)(1 - f_1 + x)$$
$$+ k(1 - \alpha)C_{A_0}^2(1 - f_2)(1 - f_2 + x) \tag{4-5}$$

Now, let us mix the two fractions, so the resulting concentrations are

$$C_A = C_{A_0}[\alpha(1 - f_1) + (1 - \alpha)(1 - f_2)]$$
$$C_B = C_{A_0}[\alpha(1 - f_1 + x) + (1 - \alpha)(1 - f_2 + x)] \tag{4-6}$$

and the resulting rate is

$$r_2 = kC_{A_0}^2[\alpha(1 - f_1) + (1 - \alpha)(1 - f_2)]$$
$$\cdot [\alpha(1 - f_1 + x) + (1 - \alpha)(1 - f_2 + x)] \tag{4-7}$$

These two rates are not the same. Comparing equations (4-5) and (4-7) gives

$$r_2 = r_1 - \alpha(1 - \alpha)(f_1 - f_2)^2 \tag{4-8}$$

and, since $\alpha < 1$, $r_2 < r_1$, the effect of segregation is to increase the point rate of this second-order reaction.

There are two important points illustrated by this example. First is the obvious one that the extent of micromixing can affect the rate of reaction through its influence on the local concentration of reactants. Second is that the magnitude of such effects also is affected by the degree of macromixing. Recall that the two volume portions had differing degrees of conversion of

A; if segregation of the two volumes had existed from the beginning of the reaction, differing conversions could only be the result of differing times of reaction for the two elements or, in present terminology, differing ages of the two. This is determined by macromixing in the system. Thus, we see that while physically the two types of mixing are independent processes, there can be interaction between them which is apparent from the point of rates of reaction.

The example can be generalized for an nth-order reaction. We let r_0 be the initial rate of reaction for a mixture of reactants in stoichiometric proportion, so that

$$r = r_0(1 - f)^n$$

when fraction f of the reactants has disappeared. Again consider two portions with f_1 and f_2 and volumes in the ratio α and $(1 - \alpha)$. Rates before and after mixing are

$$\frac{r_1}{r_0} = \alpha(1 - f_1)^n + (1 - \alpha)(1 - f_2)^n$$

$$\frac{r_2}{r_0} = [\alpha(1 - f_1) + (1 - \alpha)(1 - f_2)]^n \qquad (4\text{-}9)$$

Let α and f_1 be held constant and define

$$\gamma \equiv \frac{1 - f_2}{1 - f_1}$$

Then

$$\frac{dr_1}{dr_2} = \frac{dr_1/d\gamma}{dr_2/d\gamma} = \left[\frac{\gamma}{\alpha + (1 - \alpha)\gamma}\right]^{n-1} \qquad (4\text{-}10)$$

which is > 0 when $\gamma > 0$ and $\alpha < 1$. Also, the second derivative,

$$\frac{d^2 r_1}{dr_2^2} = \frac{d(dr_1/dr_2)/d\gamma}{dr_2/d\gamma}$$

$$= \frac{n - 1}{n} \frac{\alpha\gamma^{n-2}}{(1 - f_1)^n(1 - \alpha)[\alpha + (1 - \alpha)\gamma]^{2n-1}} \qquad (4\text{-}11)$$

is > 0 if $n > 1$ and < 0 if $n < 1$. Two additional properties of these equations are of interest:

$$\gamma = 1, \quad r_1 = r_2, \quad \text{and} \quad \frac{dr_1}{dr_2} = 1$$

$$\gamma = 0, \quad \frac{r_1}{r_2} = \alpha^{1-n} \qquad (4\text{-}12)$$

$$\alpha^{1-n} > 1 \qquad \text{for } n > 1$$

$$\alpha^{1-n} < 1 \qquad \text{for } n < 1$$

A graphical summary of the requirements attached to equations (4-10) to (4-12) is given in Figure 4.1 in terms of the rates r_1 and r_2. The requirement

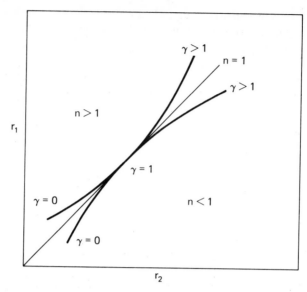

Figure 4.1 Effect of segregation on rates of nth-order reactions (after P. V. Danckwerts, *Chem. Eng. Sci.*, **8**, 93 (1958); with permission of Pergamon Press, Ltd., London)

from equation (4-12) establishes for γ of 0 or 1 and $n = 1$ that the two rates be equal. Further, $dr_1/dr_2 = 1$ for $\gamma = 1$, and $(d^2r_1/dr_2^2) = 0$ for $n = 1$; thus, $r_1 = r_2$ for $\gamma > 1$. This establishes the equality of rates for all values of γ when $n = 1$; in other words, *the degree of micromixing has no effect on the rate of a first-order reaction*. Similar reasoning leads to the conclusions that $r_1 > r_2$ for $n > 1$, and $r_1 < r_2$ for $n < 1$ save at the point $\gamma = 1$, where $r_1 = r_2$. The magnitude of the difference in rates depends on γ, which one can see is a function of the degree of macromixing since age distribution effects are contained in the differing conversions f_1 and f_2, and α, the micromixing parameter.

As we have mentioned in the introduction to this chapter, quantitative measures for degrees of micromixing intermediate to the extremes of maximum mixedness and complete segregation are difficult to establish on an a priori basis. Thus, aside from the illustration given here demonstrating the potential effects of micromixing on various types of reaction, a more quantitative treatment will depend on the various models that we can develop.

4.2 Age Distributions and Macromixing

Macromixing is concerned only with the history of individual point-volume elements during their residence in a given system. For example, consider the simple case of laminar flow through a length of cylindrical tube. We know that when the laminar flow is fully developed, there exists a parabolic profile of velocities in the direction of flow across the radius of the tube—maximum velocity at the center and zero at the wall. Hence, a point element with radial position near the wall will require a larger amount of time to traverse a given length of tubing than will a point element near the center. If the flow is segregated, which is a reasonable assumption for laminar flow at normal temperatures, there will exist a distribution of times required for elements at various radial positions to traverse the given length. The time required is called the *residence time* and the distribution the *residence-time distribution* (RTD). In this case, owing to the existence of segregation at the micromixing level, the RTD is a direct measure of the extent of macromixing in the system.

Now consider a second example, that of turbulent flow through the tube. Again there exists some sort of velocity profile, but in addition there is some level of micromixing due to eddy motion between the elements. In this case it is possible to identify a distribution of residence times for individual molecules within the system, not volume elements, and the RTD is a measure of the combined effects of macro- and micromixing.

In the most general case, then, measurements of residence-time distribution by the methods to be discussed here cannot define micromixing. In spite of this limitation, the residence-time distribution is most valuable because it provides us with as much information on the state of mixing as we can obtain short of measurements on the microscopic level.

In addition to the residence-time distribution, there are a number of other age distributions which are of importance. The distribution of ages of elements or molecules within the system is called the *internal-age distribution*, and the distribution of ages on exit from the system is the *exit-age distribution*. Finally, corresponding to the internal age distribution there is a residual lifetime or *life-expectancy distribution*, which is a measure of time remaining before exit from the system. We shall be most concerned here with the residence-time, internal-, and exit-age distributions.

It is most convenient to define these distributions quantitatively within the context of the experimental methods normally used for the determination of the residence-time distribution. Once again we base the vocabulary and the development on the pioneering work of Danckwerts [P. V. Danckwerts, *Chem. Eng. Sci.*, **2**, 1 (1953)].

Consider, then, the simple experiment shown in Figure 4.2. Fluid is flowing through a vessel of volume $\bar{V}$ at a steady-state volumetric flow rate of v. Immediately at the entrance to the vessel there is means for injecting a trace substance at a certain concentration level, C_0, and immediately at the outlet a means for detecting instantaneously and continuously the concentration of this tracer leaving the vessel. Introduction of the tracer does not affect the steady-state flow rate. The *mean residence time* in the vessel, which is the average over all elements of the fluid, is given by

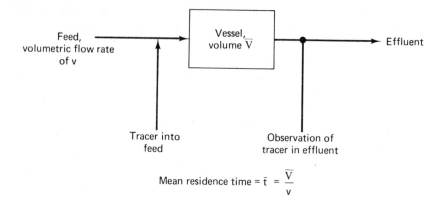

Figure 4.2 Experiment for measurement of residence time distribution

$$\bar{t} = \frac{\bar{V}}{v} \qquad (4\text{-}13)$$

Now let us perform the following experiment. Initially, feed is flowing through the vessel at steady state and no tracer is introduced. Then, at a set time, a step change in the concentration of tracer from zero to C_0 is introduced into the feed and is subsequently maintained at that level, while simultaneously the concentration of tracer in the effluent is observed and recorded. What sort of response might one observe in the concentration of tracer at the exit to this step-function change? Three plausible results are illustrated in Figure 4.3, termed an *F-diagram* by Danckwerts, in terms of the variables

$$F(t) = \frac{C_{\text{exit}}}{C_0} \qquad (4\text{-}14)$$

and

$$\theta = \frac{t}{\bar{t}} \qquad (4\text{-}15)$$

where $F(t)$, the concentration breakthrough, is the residence-time distribution and θ is the number of residence times. In Figure 4.3a, we see that all tracer molecules exit from the vessel at the same time, therefore they have identical exit ages. In Figure 4.3b, there is some distribution of ages about $\theta = 1$, with some molecules exiting before one residence time and some after, while in Figure 4.3c some molecules exit immediately upon introduction of the tracer and others remain for times $\theta \gg 1$. One might infer correctly from the shapes of the responses in Figure 4.3 that (a) and (c) represent some sort of limiting cases of mixing behavior, with (b) intermediate. These limits we will discuss in a moment, but first let us determine what relationships exist between such an experimentally determined response and the various age distributions mentioned previously. Make the following definitions:

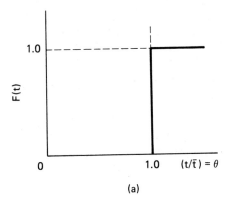

(a)

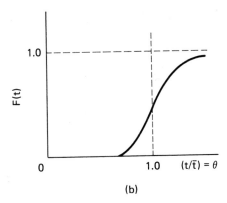

(b)

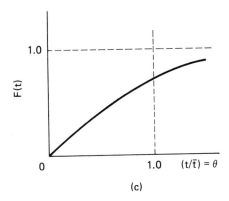

(c)

Figure 4.3 Plausible responses to a step function introduction of tracer into a flow system

1. Elements of material in vessel for t have the "age" t.
2. The fraction of material with ages between t and $t + dt$ in the system is $I(t)\, dt$.
3. The fraction of material with ages between t and $t + dt$ on leaving the system is $E(t)\, dt$.

$I(t)$ and $E(t)$ are the internal- and exit-age distribution functions, respectively. Now let us write a material balance on the tracer for a step change in its concentration from zero to 1 (the arbitrary concentration units will not affect the results here) at $t = 0$. After time t we have

$$\text{entered system} = vt$$

$$\text{still in system} = \bar{V} \int_0^t I(t')\, dt'$$

$$\text{left system} = v \int_{t''=0}^{t} \int_{t'=0}^{t''} E(t')\, dt'\, dt''$$

The last term comes from the fact that the rate of flow of tracer out of the system at any time t'' after the concentration step is given by the integral of the exit-age distribution for all ages up to t''; that is,

$$\text{flow out at } t'' = v \int_{t'=0}^{t''} E(t')\, dt'$$

The material balance is

$$\theta = \frac{vt}{\bar{V}} = \int_0^t I(t')\, dt' + \frac{v}{\bar{V}} \int_{t''=0}^{t} \int_{t'=0}^{t''} E(t')\, dt'\, dt'' \qquad (4\text{-}16)$$

The residence-time distribution, $F(t)$, including the contributions of all elements or molecules up to the time considered, must then be the time integral of the exit-age distribution:

$$F(t) = \int_0^t E(t')\, dt' \qquad (4\text{-}17)$$

Taking the time derivative of equation (4-16), we have

$$1 - \int_0^t E(t')\, dt' = \frac{\bar{V}}{v} I(t) \qquad (4\text{-}18)$$

and from (4-17), then

$$1 - F(t) = \frac{\bar{V}}{v} I(t) \qquad (4\text{-}19)$$

which relates the measured residence time distribution $F(t)$ to the internal age distribution, $I(t)$. Further, from the derivative of equation (4-17) we have

$$\frac{d}{dt}[F(t)] = E(t) \qquad (4\text{-}20)$$

It is also convenient to remember from the definitions of internal- and exit-age distributions that

$$\int_0^\infty I(t)\, dt = \int_0^\infty E(t)\, dt = 1 \qquad (4\text{-}21)$$

since the integral for all ages from 0 to ∞ represents the total amount of material (fraction $= 1$) in the system.

A second type of experiment often used in the determination of RTD is the response to a pulse input of tracer rather than a step function. Here a total quantity, Q, of the tracer is injected into the feed stream at a concentration of C_0 over a small time period Δt. Differing residence times of molecules in the system will lead to a dispersion of the pulse with typical response curves shown in the *C-diagram* of Figure 4.4, corresponding to those illustrated in Figure 4.3. The response of the $C(t)$ curves of Figure 4.4 is just the derivative of the $F(t)$ curves in Figure 4.3.

$$\frac{d}{dt}[F(t)] = E(t) \qquad (4\text{-}22)$$

where

$$E(t) = \frac{C(t)}{Q} \qquad (4\text{-}23)$$

Thus, either pulse-response or step-function-response experiments give sufficient information to permit evaluation of exit-age, internal-age, and residence-time distributions. The average age or mean residence time, which we have defined intuitively in equation (4-13), can be more precisely derived in terms of the time average of the exit-age distribution:

$$\bar{t} = \frac{\int_0^\infty t E(t)\, dt}{\int_0^\infty E(t)\, dt} = \int_0^\infty t E(t)\, dt \qquad (4\text{-}24)$$

From equation (4-20) we may write this as

$$\frac{v\bar{t}}{\bar{V}} = \int_0^1 \frac{vt}{\bar{V}}\, d[F(t)] = 1 \qquad (4\text{-}25)$$

so that

$$\bar{t} = \frac{\bar{V}}{v}$$

The average age of material in the system at any time $\bar{t}_1$ is

$$\bar{t}_1 = \int_0^\infty t I(t)\, dt = \frac{v}{2\bar{V}} \int_0^1 t^2\, d[F(t)] \qquad (4\text{-}26)$$

This is most easily evaluated graphically or numerically from the $F(t)$ diagram.

Pulse-response and step-function-response experiments are perhaps the easiest to carry out and to analyze; however, any perturbation-response technique can be used to determine age distributions. Kramers and Alberda

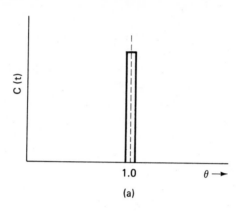

(a)

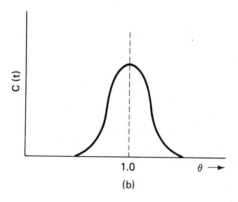

(b)

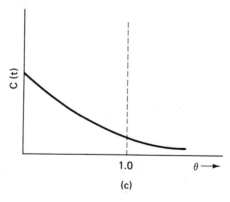

(c)

Figure 4.4 Responses to a pulse input of tracer into a flow system

194

[H. Kramers and G. Alberda, *Chem. Eng. Sci.*, **2** 173 (1953)] describe a
frequency-response analysis, and a general treatment of arbitrary input functions, errors associated with input and measurement, and other factors is
given in the review by Levenspiel and Bischoff [O. Levenspiel and K. B.
Bischoff, *Advan. Chem. Engr.*, **4**, 95 (1963)].

While the concepts involved in the definition of these distributions are
relatively simple ones, practical application is sometimes confusing because
of the differing time measures used. There are two time variables and one
time constant that we must consider:

1. t = real time, seconds, minutes, and so on. This is the time variable
   employed in the age-distribution definitions up to this point.
2. $\bar{t}$ = the nominal residence time in the system, also seconds, minutes,
   and so on, defined by equation (4-13).
3. θ = the number of residence times, $t/\bar{t}$, a dimensionless quantity.

Experimental data on exit-age or residence-time distribution most often
take the form of discrete values of tracer concentration measured at successive
time intervals after introduction of the tracer. Thus, the integrals involved
are normally replaced by summations in the analysis of actual data. We will
illustrate here the procedure for analysis of a pulse-response experiment.
Available are tracer concentrations in the effluent, $C(t)$ and corresponding
times, and from these data we would like to determine the exit-age distribution, or $E(\theta)\, d\theta$, the distribution in terms of the residence-time variable θ.
First determine $E(t)$ from $C(t)$ versus t by

$$E(t) = \frac{C(t)}{\sum C(t)\,\Delta t} = \frac{C(t)}{Q} \tag{4-27}$$

Division by $Q = \sum C(t)\,\Delta t$ in equation (4-27) ensures the normalization of
the distribution required by equation (4-21); graphically, the area under a
plot of $E(t)$ versus t must be unity. Next, determine the value of $\bar{t}$. This can be
done in either of two ways:

$$\bar{t} = \frac{\sum tC(t)\,\Delta t}{\sum C(t)\,\Delta t} \tag{4-28}$$

or

$$\bar{t} = \sum tE(t)\,\Delta t \tag{4-29}$$

Now, with values for $E(t)$ and t determined, $E(\theta)$ can be evaluated from the
simple relationship

$$E(\theta) = \bar{t}E(t) = \frac{\sum tC(t)\,\Delta t}{\sum C(t)\,\Delta t} \frac{C(t)}{\sum C(t)\,\Delta t} \tag{4-30}[1]$$

It will be seen later that it is the exit-age distribution (or the derivative of the

---

[1]Equation (4-30) and corresponding relationships for other distributions in real and
dimensionless time variables are obtained from the following:

F curve) that is of importance in determining the effects of nonideal flow on actual chemical reactor performance.

Let us return finally to a more thorough analysis of the limiting types of response indicated by Figures 4.3 and 4.4a and c. In case (a), all molecules have an identical residence time in the system and the residence time distribution is given by

$$F(t) = 0 \qquad t < \bar{t}$$
$$F(t) = 1 \qquad t > \bar{t} \tag{4-31a}$$

or

$$F(\theta) = 0 \qquad \theta < 1$$
$$F(\theta) = 1 \qquad \theta > 1 \tag{4-31b}$$

This type of RTD is normally associated with the terms "plug flow" or (misleading) "no mixing," since the characteristics of plug flow through a tube—with uniform velocity across the radius and no mixing in the direction of flow—require a unique residence time. Once again, let it be stated that such response does not define the state of micromixing.

The response in case (c) is at the opposite end of the spectrum from plug flow and is the limit established when the internal-age distribution and the external-age distribution are identical. The consequence of this is an exponential RTD, as shown by the following. Rewrite the material balance, equation (4-16), subject to

$$\int_0^t E(t')\, dt' = \int_0^t I(t')\, dt' = F(t) \tag{4-32}$$

Then

$$\frac{vt}{V} = F(t) + \frac{v}{V} \int_{t''=0}^t F(t'')\, dt'' \tag{4-33}$$

Taking the derivative with respect to time and rearranging yields

$$\frac{d[F(t)]}{dt} + \frac{v}{V} F(t) = \frac{v}{V} \tag{4-34}$$

$$\int_0^\infty E(t)\, dt = \int_0^\infty E(\theta)\, d\theta = 1$$

and

$$\theta = \frac{t}{\bar{t}}$$

$$dt = \bar{t}\, d\theta$$

Then

$$\int_0^\infty E(\theta)\, d\theta = \int_0^\infty \bar{t} E(t)\, d\theta$$

so

$$\bar{t} E(t) = E(\theta)$$

Solving for $F(t)$,

$$\int_0^{F(t)} d[F(t)] = e^{-(v/V)t} \int_0^t \frac{v}{V} e^{(v/V)t} \, dt$$

$$F(t) = 1 - e^{-(v/V)t} \tag{4-35a}$$

$$F(\theta) = 1 - e^{-\theta} \tag{4-35b}$$

The corresponding exit-age distribution is

$$E(t) = \frac{d[F(t)]}{dt} = \frac{v}{V} e^{-(v/V)t} \tag{4-36a}$$

or

$$E(\theta) = e^{-\theta} \tag{4-36b}$$

The exponential RTD is associated with the term "perfect mixing," although this may or may not be the case on both the macroscopic and microscopic levels.

A summary of the possible limiting combinations of macro- and micro-mixing, and the result in terms of a residence-time distribution, is given in Figure 4.5.

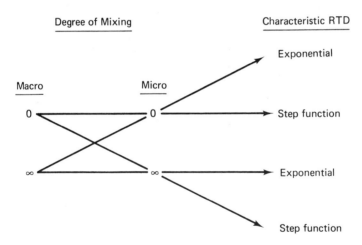

Figure 4.5 Residence-time distributions for limiting combinations of macro- and micromixing

4.3   Mixing Models—Reactors with Ideal Flow

The step-function and exponential-residence-time distributions of Figure 4.5 can be modeled by two different types of flow systems. For the step-function response we have already alluded to the model of plug flow through a tube, which is, indeed, a standard model for this response. The exponential response, described previously as the result of the equality of internal- and exit-age distributions, requires a bit more thought. In the following we will derive the equations for the mixing models and then the corresponding reactor models for these two ideal limits.

a. Plug flow—the mixing model

Consider a differential length, dz, of cylindrical conduit as shown in Figure 4.6. Fluid with a uniform velocity u in the axial direction is passing

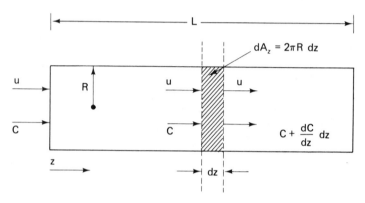

Figure 4.6 Plug-flow model

through the differential volume $A\,dz$, where A is the cross section of the tube. At a time $t = 0$, a step input of tracer is introduced uniformly across the cross section at the entrance to the tube at concentration C_0. We write the general unsteady-state mass balance for the differential volume in the normal fashion:

$$\text{input} - \text{output} = \text{accumulation}$$

where

$$\text{tracer input} = uCA$$

$$\text{tracer output} = uA\left(C + \frac{dC}{dz} \cdot dz\right)$$

$$\text{tracer accumulation} = A\,dz\,\frac{dC}{dz}$$

Equating the terms,

$$uCA - uA\left(C + \frac{dC}{dz} \cdot dz\right) = A\,dz\,\frac{dC}{dt}$$

or

$$-u\frac{dC}{dz} = \frac{dC}{dt} \tag{4-37}$$

with the initial and boundary conditions

$$t < 0, \quad z = 0, \quad C = 0$$
$$t > 0, \quad z = 0, \quad C = C_0$$
$$t = 0, \quad z > 0, \quad C = 0$$
$$t = 0, \quad z < 0, \quad C = C_0$$

The solution of equation (4-37) for the tracer concentration at the exit $(z = L)$ is just

$$C = 0, \quad t < \frac{L}{u}$$

$$\text{(4-38)}$$

$$C = C_0, \quad t > \frac{L}{u}$$

which is the step-function response of Figure 4.3.

The specific mixing assumptions involved in this model are the absence of mixing in the axial direction, and uniformity of concentration and velocity in the radial direction.

b. Plug flow—the reactor model

We will make several modifications to the plug-flow mixing model in order to derive the corresponding reactor model. First, since our primary interest will be in the steady-state behavior of the reactor, the time-dependent accumulation term need not be included and the corresponding initial condition disappears. Second, since there is now a chemical reaction occuring in the reactor, a term expressing the disappearance of reactants (or appearance of products) due to reaction must be added. Third, we are not dealing with a tracer experiment here, so the concentration C refers to that of the reactant or product to which the chemical rate term pertains. Thus,

$$\text{input} = uCA$$

$$\text{output} = uA\left(C + \frac{dC}{dz} \cdot dz\right) + (-r)A \, dz$$

where

$$C = \text{concentration of reactant or product}$$

$$(-r) = \text{rate of reaction per unit volume}$$

Equating input and output for steady-state operation:

$$uCA - UA\left(C + \frac{dC}{dz} \cdot dz\right) - (-r)A \, dz = 0$$

or

$$-u\frac{dC}{dz} = -r \tag{4-39}$$

The rate of reaction term, $(-r)$, as employed in equation (4-39), is a net positive quantity if C refers to reactant, negative if D refers to product. The boundary conditions define $C = C_0$ (reactant) at $z = 0$. The rate of reaction is a function of C, so it is convenient to rearrange equation (4-39) to

$$\frac{dC}{(-r)} = -\frac{dz}{u} \tag{4-40}$$

which can be integrated to

$$\int_{C_0}^{C} \frac{dC}{(-r)} = -\frac{L}{u} \tag{4-41}$$

The integrated form of the left-hand side of this equation of course depends on the form of rate equation expressing $(-r)$. Equation (4-41) is the basic design equation of the *plug-flow reactor* (PFR) model and may be encountered in a number of different forms. The most familiar of these employs the reactor volume, $\bar{V}$, instead of length; total mass or molal flow rate, F; and conversion, x. This is commonly written

$$\int_{x_0}^{x} \frac{dx}{(-r)} = \frac{\bar{V}}{F} \tag{4-42}$$

Remember that the rate equation must be consistent with the measure of the extent of reaction employed; if conversion is used, $(-r)$ must be expressed in terms of x; if concentration is used, $(-r)$ is in terms of C. If F is mass flow rate, $(-r)$ must be in mass units; if F is molal flow rate, $(-r)$ must be in molal units. Equation (4-42) is also often used as a pseudo-homogeneous model for PFR catalytic reactors. In such cases the rate constant contained in $(-r)$ would be expressed in terms per volume or weight of catalyst. In the former case, $\bar{V}$ would refer to catalyst volume and a value for bed porosity would be required to obtain reactor volume. In the latter case, catalyst weight would be obtained, and the catalyst bulk density would be required to obtain $\bar{V}$. Finally, it is important to realize that F and $(-r)$ are keyed to each other not only as to dimension but also as to component. Thus, if a mixture of two reactants is fed to the reactor, $(-r)$ represents the rate of reaction in terms of one of the reactants and F the feed rate of *that* reactant, not the total feed rate.

If the feed rate is converted to volumetric units, the inverse of the ratio on the right-hand side of equation (4-42)

$$\frac{vF}{\bar{V}} = \frac{v}{\bar{V}} = SV \tag{4-43}$$

is the space velocity, SV, with units of reciprocal time. It is a direct measure of the maximum feed rate which can be used per unit volume of reactor in order that the stated conversion level x can be attained. If the feed consists of material in addition to that for which the reactor mass balance is written, the total feed rate and corresponding density must be used in equation (4-43). Small space velocities mean slow reactions and a correspondingly large reactor volume required to attain a specified conversion.

The reciprocal of the space velocity, the space time, is sometimes employed. It can be misleading, however, since the space time is in general equal to the actual residence or contact time within the reactor only if the temperature, pressure, and reaction mixture density are constant throughout the reactor. A generalized expression for the residence time in a PFR can be derived as follows in terms of molar flow rates:

$$dt = \frac{d\bar{V}}{Fv(z)}$$

where the total volumetric flow rate $Fv(z)$ may vary along the reactor length (but not across the radius in the PFR model), owing to change in moles on reaction or variations in temperature or pressure. In this formulation, then, $v(z)$ is the specific volume of reaction mixture based on an inlet total feed rate of F moles/time. If we substitute for $d\bar{V}$ from the derivative of equation (4-42),

$$d\bar{V} = \frac{F\,dx}{(-r)}$$

and

$$dt = \frac{dx}{(-r)v(z)} \tag{4-44}$$

The true residence time is then

$$t_R = \int_0^x \frac{dx}{(-r)v(z)} \tag{4-45}$$

c. Perfect mixing—the mixing model

The equality of internal- and exit-age distributions for this case implies that each molecule within the system at a given instant has a past history indistinguishable from any other. Its life expectancy within the system is thus independent of when it entered the system, and all molecules or elements have an equal chance of leaving. This can only be so if there is a uniform concentration within the system, which is equal to the concentration at the exit. A working model for this situation is shown in Figure 4.7, consisting of a very well agitated vessel of volume $\bar{V}$ with a steady volumetric flow rate v. At time $t = 0$ a step input of tracer at concentration C_0 is introduced to the system. The unsteady-state mass balance on tracer is

$$\text{tracer input} = vC_0$$

$$\text{tracer output} = vC$$

$$\text{tracer accumulation} = \bar{V}\frac{dC}{dt}$$

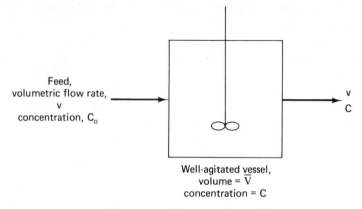

Figure 4.7 Perfect mixing model

where C is the uniform tracer concentration within the vessel. Equating the terms,

$$\bar{V}\frac{dC}{dt} = v(C_0 - C)$$

or

$$\frac{dC}{dt} + \frac{1}{\bar{t}}C = \frac{1}{\bar{t}}C_0 \qquad\qquad (4\text{-}46)$$

where $\bar{t}$ is the residence time ($\bar{V}/v$) and the limits of integration are $0 \leq t' \leq t$ and $0 \leq C' \leq C$. Carrying out this integration we obtain

$$C = C_0(1 - e^{-t/\bar{t}}) \qquad\qquad (4\text{-}47)$$

which is the exponential response of Figure 4.3. The specific mixing assumptions involved in this model are the spatial uniformity of concentration within the vessel volume, and the equality of exit and internal concentrations.

d. Perfect mixing—the reactor model

Again we make several modifications to the mixing model to obtain the reactor model. These are the assumption of steady state, the inclusion of a reaction term in the mass balance, and the reference of concentration to that of a reactant or product corresponding to the rate term rather than that of a tracer. The mass balance now becomes

$$\text{input} = vC_0$$

$$\text{output} = vC + (-r)\bar{V}$$

where the rate of reaction per unit volume, $(-r)$, is defined as for the PFR. The result is very simple, since the assumption of concentration uniformity results in an algebraic relationship between inlet and exit compositions:

$$vC_0 = vC + (-r)\bar{V}$$

or

$$C = C_0 - (-r)\bar{t} \tag{4-48}$$

Equation (4-48) is the basic design equation for what is popularly called a *continuously stirred tank reactor* (CSTR). The derivation assumes equality of volumetric flow rate of feed and effluent; as in the case of the PFR, the residence-time definition must be changed if this is not so. In most applications of the CSTR, however, reactions in the liquid phase are involved and volume changes on reaction are not important.

e. Some additional limiting reactor models

The PFR and CSTR models encompass the extremes of the residence-time distributions shown in Figure 4.3; however, the batch reactor and the laminar flow reactor, both of which we have already mentioned in this chapter, are also types exhibiting a well-defined mixing behavior. The batch reactor is simply represented by the perfect mixing model with no flow into or out of the system, and has been extensively treated in chapter 1.

The laminar flow reactor with segregation and negligible molecular diffusion has a residence-time distribution which is the direct result of the velocity profile in the direction of flow of elements within the reactor. To derive the mixing model of this reactor, let us start with definition of the velocity profile:

$$u(r_p) = \frac{2v}{\pi R^2}\left[1 - \left(\frac{r_p}{R_0}\right)^2\right] \tag{4-49}$$

where R_0 is the radius of the tube, r_p the radial position at which the velocity $u(r_p)$ is determined, and v again the volumetric flow rate. For length L of tubing the residence time $t(r_p)$ corresponding to $u(r_p)$ is

$$t(r_p) = \frac{L}{u(r_p)} \tag{4-50}$$

and the average residence time $\bar{t}$ is given by

$$\bar{t} = \frac{L}{\bar{u}}$$

where $\bar{u}$ is the average velocity, $\bar{u} = v/\pi R_0^2$. Solving equations (4-49) and (4-50) for r_p gives

$$r_p = R_0\sqrt{1 - L/\bar{u}t} = R_0\sqrt{1 - \bar{t}/t}$$

Now

$$F(t) = \frac{C}{C_0} = \frac{\int_0^{r_p} 2\bar{u}[1 - (r_p/R_0)^2]C \cdot 2\pi r_p \, dr_p}{\int_0^{R_0} \bar{u}C 2\pi r_p \, dr_p} \tag{4-51}$$

or

$$F(t) = \frac{1}{\pi R_0^2 \bar{u}} \int_0^{\sqrt{1-\bar{t}/t}} 2u[1 - (r_p/R_0)^2]2\pi r_p \, dr_p$$

Evaluation of this integral yields

$$F(t) = 1 - \frac{\bar{t}^2}{4t^2} \tag{4-52}$$

and the exit-age distribution is

$$E(t) = \frac{dF(t)}{dt} = \frac{\bar{t}^2}{2t^3} \tag{4-53}$$

A comparison of the laminar flow residence time distribution with corresponding plug flow and perfect mixing results is shown in Figure 4.8.

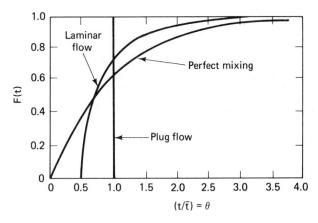

Figure 4.8 Comparison of plug flow, perfect mixing, and laminar flow residence-time distributions

The reactor model for laminar flow is similar to that for plug flow, except for the radial dependence of velocity. The analog of equation (4-39) is

$$-u(r_p)\frac{dC(r_p)}{dz} = (-r)$$

or

$$\int_{C_0}^C \frac{dC(r_p)}{(-r)} = -\frac{L}{u(r_p)} = -\frac{\pi R^2 L}{2v[1 - (r_p/R_0)^2]} = \int_0^x \frac{dx(r_p)}{(-r)} \tag{4-54}$$

$$= -\frac{\bar{V}}{2v[1 - (r_p/R_0)^2]}$$

Solution of equation (4-54) explicitly for exit concentration depends on the form of the rate law. If an explicit solution is obtained the value of $C(r_p)$ must be averaged over the exit cross section to determine the average exit concentration, $\langle C \rangle$:

$$\langle C \rangle = \frac{\int_0^R u(r_p)C(r_p)r_p\,dr_p}{\int_0^R u(r_p)r_p\,dr_p} \tag{4-55}$$

4.4   Applications of Ideal Reactor Models

In each case of section 4.3 the reactor models obtained on the basis of the mixing models were left in general form, with specific results for concentration or conversion versus reactor size depending on the form of the rate law involved. In the following discussion we shall derive a number of expressions for particular rate laws which relate conversion, reactor size, and residence time for homogeneous, isothermal ideal reactors.

a. Plug-flow reactors

For the plug-flow reactor there is a formal analogy with the batch reactor equations derived in chapter 1 if there is no volume expansion or contraction on reaction. As an example, consider the PFR design equation, (4-42), written for first-order irreversible kinetics (A $\rightarrow$ B):

$$\frac{\bar{V}}{F_A^\circ} = \int_0^x \frac{dx}{(-r)} \tag{4-42a}$$

$$-r = kC_A$$

where F_A° is the entering molal feed rate of A, no B is present in the feed, $-r$ is the rate of reaction in moles A/time-volume, and x is the conversion of A defined analogously to that of chapter 1 as the ratio of moles reacted to moles fed. Hence,

$$x = \frac{F_A^\circ - F_A}{F_A^\circ} = 1 - \frac{F_A}{F_A^\circ} \tag{4-56}$$

Now

$$F_A = F_A^\circ(1 - x)$$

and the (constant) total molal flow is

$$F_T = F_A + F_B = F_A^\circ(1 - x) + F_A^\circ x = F_A^\circ$$

If the specific volume of the reaction mixture is v, the reactant concentration at any point can be written in terms of the conversion:

$$C_A = \frac{F_A}{F_A^\circ v} = \frac{F_A^\circ(1 - x)}{F_A^\circ v} = \frac{1 - x}{v} \tag{4-57}$$

and

$$(-r) = \frac{k(1 - x)}{v} \tag{4-58}$$

Substitution into the PFR equation gives

$$\frac{\bar{V}}{F_A^\circ} = \int_0^x \frac{v\,dx}{k(1-x)}$$

which integrates directly to

$$\frac{\bar{V}}{F_A^\circ v} = -\frac{1}{k}\ell n\,(1-x) \tag{4-59}$$

The left side of equation (4-59) is just the residence time, t_R, and could be obtained directly from the definition of equation (4-45) for constant v. Thus,

$$t_R = -\frac{1}{k}\ell n\,(1-x) \tag{4-60}$$

which is identical to the result given in equation (1-48) for this rate law in a homogeneous batch reactor.

Corresponding derivations for the second-order irreversible cases $2A \rightarrow C + D$ or $A + B \rightarrow C + D$ yield (assuming equimolal amounts of A and B in the feed in the second case)

$$t_R = \frac{v}{k} \cdot \frac{x}{1-x} \tag{4-61}$$

which is again directly comparable to the batch results in chapter 1.

Volume changes on reaction can be accounted for by writing explicitly the dependence of concentration on conversion. As an example, consider the first-order reaction $A \rightarrow C + D$. Here total moles are

$$F_T = F_A^\circ(1-x) + 2F_A^\circ x = F_A^\circ(1+x)$$

assuming no C or D in the feed. The corresponding concentration of reactant A is

$$C_A = \frac{F_A^\circ(1-x)}{F_A^\circ(1+x)v}$$

and

$$\int_0^x \frac{dx}{(-r)} = \int_0^x \frac{v(1+x)\,dx}{k(1-x)} \tag{4-62}$$

giving

$$t_R = \frac{1}{k}[-2\ell n\,(1-x) - x] \tag{4-63}$$

General expressions for PRF $t_R - x$ relationships, in terms of the volume change factor ϵ defined in chapter 1, are summarized in Table 4.1 for a number of rate expressions.

Plug-flow reactors are often employed in the laboratory for the measurement of reaction rates. Most often such application is made for heterogeneous gas/solid catalytic reactions; however, the details of the procedure do not differ substantially between homogeneous and heterogeneous systems. The

TABLE 4.1

PLUG-FLOW REACTOR CONVERSION/RESIDENCE-TIME RELATIONSHIPS

1. nth-order irreversible reaction: $A \longrightarrow \nu$ products

$$t_R = \frac{1}{\upsilon} \int_0^x \frac{dx}{kC_A^n} = \frac{\upsilon^{n-1}}{k} \int_0^x \frac{(1 + \epsilon x)^n \, dx}{(1 - x)^n}$$

2. Zero-order irreversible reaction:

$$t_R = \frac{1}{\upsilon} \int_0^x \frac{dx}{k} = \frac{x}{k\upsilon}$$

3. First-order irreversible reaction:

$$t_R = \frac{1}{\upsilon} \int_0^x \frac{dx}{kC_A} = -\frac{1 + \epsilon}{k} \ell n \, (1 - x) - \frac{\epsilon x}{k}$$

4. Second-order irreversible reactions:

$$t_R = \frac{1}{\upsilon} \int_0^x \frac{dx}{kC_A^2} \text{ or } \frac{1}{\upsilon} \int_0^x \frac{dx}{kC_A C_B} \qquad (C_{A_0} = C_{B_0})$$

$$= \frac{2\upsilon\epsilon}{k}(1 + \epsilon) \ell n \, (1 - x) + \frac{\upsilon\epsilon^2 x}{k} + \frac{\upsilon(\epsilon + 1)^2}{k}\left(\frac{x}{1 - x}\right)$$

5. First-order reversible reaction: $A \rightleftharpoons \nu B$ (k_1 forward, k_2 reverse)

$$t_R = \frac{\epsilon x}{k_1 \alpha} - \frac{\alpha + \epsilon}{k_1 \alpha^2} \ell n \, (1 - \alpha x)$$

$$\alpha = 1 + \frac{k_2}{k_1}(1 + \epsilon)$$

6. General nth-order irreversible reaction with no volume change:

$$1 - x = [1 + (n - 1)C_0^{n-1} k t_R]^{1/(1-n)}$$

experiments normally consist of measuring in a series of runs the conversion of reactant (and product distribution in complex reactions) as a function of residence time or space velocity in the reactor. Each run is made at constant temperature, total pressure or composition and inlet reactants ratio, and the same temperature and pressure. Similar experiments are repeated until the entire range of temperature, composition, and pressure of interest have been investigated. If the conversions are high, conversion-time data may be interpreted in terms of relationships such as those of Table 4.1 or corresponding forms in terms of space velocity, according to the convenience of the experimenter, using graphical or numerical procedures analogous to those described in chapter 1. The PFR in such application is described as an *integral reactor*, since the measured conversions are the integral result of the reaction kinetics over a range of reaction mixture compositions. If the conversion is kept at a very low level in experimentation, dx may be approximated by the small Δx, and equation (4-42), for example, becomes

$$\frac{\Delta x}{(-r)} = \frac{\bar{V}}{F}$$

which can be solved directly for the rate of reaction:

$$(-r) = \left(\frac{F}{\bar{V}}\right) \Delta x \tag{4-64}$$

In this application, the PFR is termed a *differential reactor*.

Direct rate measurements are always preferable if one is trying to determine the form of reaction kinetics, reaction orders, rate constants, and so on, as pointed out in chapter 1. There are, however, a number of practical operational difficulties involved in utilization of differential reactor techniques which limit somewhat their utility. The primary difficulty is the precise measurement of very small differences in concentration required by the "differential" conversion level—a consideration that also involves one in the hot question of what conversion range can be used in the approximation of "differential." If the form of the rate equation has been established and one is merely trying to determine values for the rate constants, the errors involved in the differential conversion approximation may be estimated by comparison of results from the differential and integral forms of the PFR equation. A second operational difficulty, which is also involved in integral reactor applications, is the experimental inconvenience of independently varying t_R or $(\bar{V}/F)$ while maintaining convenient or required conversion levels *and* the ideal flows required for analysis of data in terms of the PFR model. This last condition refers only to integral reactor operation, for at sufficiently low conversions the effects of nonideal flows are negligible regardless of whether they arise from micro- or macromixing processes. Further problems with integral reactor operation may arise due to the appearance of gradients of temperature or pressure appearing at low or high space velocities.

When these considerations are appended to other restrictions that we shall discuss later concerning inter- and intraphase gradients of temperature and concentration in heterogeneous reactions, it is clear that successful utilization of the laboratory PFR for measurement of intrinsic kinetics is often a delicate matter. There exist a large number of "kinetic" data, particularly in the older literature, which are unreliable because of ignorance or inability on the part of the investigators to effect the experimental compromises required.

b. Continuously stirred tank reactors

The analysis of the CSTR is simpler than for the PFR, since the basic relationship between concentration or conversion and rate is an algebraic rather than a differential one. Utilization of the CSTR also offers some advantages over the PFR in terms of ease of control and steady operation resulting from the well-mixed state within the reactor. There is also a significant disadvantage, however; the reaction occurring always goes at the slowest possible rate, that is, at the rate corresponding to the concentration level of reactant in the product stream. This disadvantage can be compensated for somewhat by putting several CSTR in series rather than using a single reactor of larger volume. For this the analysis of CSTR sequences, as well as of single reactors, is important.

For a first-order, irreversible reaction $A \longrightarrow B$, equation (4-48) becomes

$$C = C_0 - kC\bar{t}$$

and solving for the exit concentration C,

$$\frac{C}{C_0} = \frac{1}{1 + k\bar{t}} \tag{4-65}$$

The substitution of $(-r) = kC$, with C the exit concentration, is the result of the mixing state within the reactor.

For second-order kinetics, $2A \longrightarrow C + D$, a similar treatment yields

$$C = C_0 - kC^2\bar{t}$$

and

$$C = \frac{-1 + \sqrt{1 + k\bar{t}C_0}}{2k\bar{t}} \tag{4-66}$$

A summary of results for several simple-order rate equations is given in Table 4.2. The effect of volume changes on reaction is presumed negligible in these results. Note that in the general nth-order case and the second-order

TABLE 4.2

CSTR CONVERSION/RESIDENCE-TIME RELATIONSHIPS—SINGLE REACTORS

1. nth-order irreversible reaction: $A \longrightarrow v$ products

$$\frac{C}{C_0} = \frac{1}{1 + kC^{n-1}\bar{t}}$$

2. Zero-order irreversible reaction:

$$C = C_0 - k\bar{t}$$

3. First-order irreversible reaction:

$$\frac{C}{C_0} = \frac{1}{1 + k\bar{t}}$$

4. Second-order irreversible reactions:

$2A \longrightarrow C + D$

$$C = \frac{-1 + \sqrt{1 + 4k\bar{t}C_0}}{2k\bar{t}}$$

$A + B \longrightarrow C + D$

$$\frac{C_A}{C_{A_0}} = \left(1 + \frac{kC_{B_0}\bar{t}}{1 + kC_A\bar{t}}\right)^{-1}$$

$$\frac{C_B}{C_{B_0}} = \frac{1}{(1 + kC_A\bar{t})}$$

5. First-order reversible reaction: $A \rightleftharpoons B$ (k_1 forward, k_2 reverse)

$$C_A = \frac{C_{A_0}}{1 + k_1\bar{t} + k_2\bar{t}} + \frac{k_2\bar{t}(C_{A_0} + C_{B_0})}{1 + k_1\bar{t} + k_2\bar{t}}$$

reaction between different reactants it is not possible to obtain an explicit solution for exit concentration. This leads to difficulty in the analysis of CSTR sequences, as seen below.

CSTR are often employed for homogeneous polymerization reactions, and owing to the special nature of the kinetics of the chain reactions involved, the analysis is changed somewhat. Let us consider the following chain for the conversion of monomer M into a series of polymeric products R_j, similar to the chain transfer reaction (XI) in chapter 1:

$$M \longrightarrow R_1 \qquad \text{(initiation, } k_i\text{)}$$

$$R_1 + M \longrightarrow R_2$$

$$\cdot$$
$$\cdot \qquad \text{(propagation, } k_p\text{)}$$
$$\cdot$$

$$R_j + M = R_{j+1}$$

The rates of initiation and propagation are, respectively,

$$-r_i = k_i C_M$$
$$-r_j = k_p C_{R_j} C_M$$

where the propagation-rate constants are the same for each step. For the monomer, the rate of reaction to be substituted into equation (4-48) is

$$-r_M = k_i C_M + k_p C_M \sum_{j=1}^{\infty} C_{R_j} \tag{4-67}$$

and the CSTR material balance is

$$\frac{C_{M_0} - C_M}{\bar{t}} = k_i C_M + k_p C_M \sum_{j=1}^{\infty} C_{R_j} \tag{4-68}$$

In order to calculate the conversion of monomer, we must determine a value for $\sum_{j=1}^{\infty} C_{R_j}$ in equation (4-68). This can be done by writing balances for each of the R_j and determining their sum. For R_1:

$$-r_1 = -k_i C_M + k_p C_{R_1} C_M \tag{4-69}$$

Substitution in equation (4-48) gives

$$k_i C_M = \frac{C_{R_1}}{\bar{t}} + k_p C_{R_1} C_M \tag{4-70}$$

and in general

$$k_p C_M C_{R_{j-1}} = \frac{C_{R_j}}{\bar{t}} + k_p C_{R_j} C_M \qquad (j = 2, \ldots, \infty) \tag{4-71}$$

Adding the balances for all individual species, equations (4-70) and (4-71) give

$$k_i C_M = \sum_{j=1}^{\infty} \frac{C_{R_j}}{\bar{t}} \tag{4-72}$$

Substituting equation (4-72) into (4-68) gives the final expression for con-

Assume further that we have available, from whatever source analytical or experimental, information on the dependence of $(-r)$ on C, as shown in Figure 4.12a. The graphical interpretation of equation (4-81) is shown in Figure 4.12b. The quantity $(-r)/(C_0 - C)$ is the tangent of the angle included between $C_0 - C$, on the abcissa, and the straight line between the locus of C_0 on the abcissa and $(-r)$, corresponding to the exit concentration C, located on the $(-r)$–C curve. Thus, $(1/\bar{t})$ is also the tangent of this angle. For the construction, we start at C_0 on the abcissa and draw a straight line at an angle $\arctan(1/\bar{t})$ from this point to the $(-r)$–C curve. The intersection defines the corresponding value of $(-r)$, and a vertical line from the intersection to the abcissa determines C. The extension of this procedure to a sequence of CSTR is easily visualized. In Figure 4.12c the method is illustrated for a sequence of 4 CSTR of differing residence times. If the sequence is not an isothermal one, a number of $(-r) - C$ curves for the individual temperature levels involved will be required.

The single CSTR is increasing in popularity as a laboratory reactor, primarily because of the development of its application to measurement of gas/solid heterogeneous catalytic kinetics as first suggested by Carberry [J. J. Carberry, *Ind. Eng. Chem.*, **56**, 39 (1964); D. J. Tajbl, J. B. Simons, and J. J. Carberry, *Ind. Eng. Chem. Fundls.*, **5**, 171 (1966)]. A number of related designs based on internal recirculation of the reaction mixture through a small fixed bed of catalyst may also be treated conceptually as CSTRs. Solving equation (4-48) for the rate,

$$(-r) = \frac{C_0 - C}{\bar{t}} \tag{4-82}$$

Measurements of inlet and outlet concentrations corresponding to the residence time $\bar{t}$ then yield rate data directly. The measured rate, of course, is that corresponding to C, and the relation between rate and concentration can be obtained experimentally over the desired range of C by variation of the residence time in the reactor. Except at very low residence times there will be substantial differences between C_0 and C, so the difficult analytical problems encountered in measuring rates with a differential PFR are, by and large, absent in the CSTR system. In addition, owing to the mixing within the CSTR, the development of gradients that obscure kinetic measurements is easily controlled. The same general experimental procedure is used with the CSTR as for the PFR. Each series of runs in which residence time is varied is carried out at constant temperature and inlet reactants ratio; then another series at a different inlet reactants ratio and the same temperature—continuing until the desired range of concentrations, conversions, and temperatures have been investigated. The advantage of the CSTR is that direct rate data are provided under all experimental conditions. Additional details on the laboratory applications of the CSTR have been reviewed in the previously cited papers by Carberry.

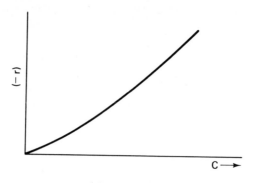

(a) Rate/Concentration
relationship

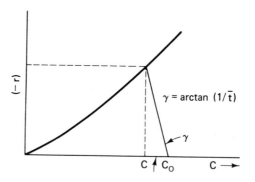

$\gamma = \arctan\ (1/\bar{t})$

γ

$C \uparrow C_0$

(b) Graphical Construction
for a Single CSTR

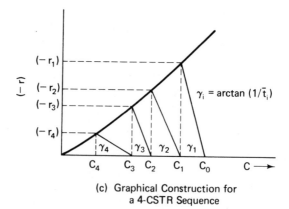

$(-r_1)$

$(-r_2)$

$(-r_3)$

$(-r_4)$

$\gamma_i = \arctan\ (1/\bar{t}_i)$

γ_4 γ_3 γ_2 γ_1

C_4 C_3 C_2 C_1 C_0

(c) Graphical Construction for
a 4-CSTR Sequence

Figure 4.12 Graphical construction for a CSTR series

218

c. Semibatch reactors

Semibatch reactors are stirred tank reactors in which there is a non-steady flow through the system. In their most normal application, treated here, one reactant will be contained initially in the tank and a second continuously added to it, with no flow out of the vessel (Figure 4.13). Other applications may involve withdrawal of a product stream but at a different rate from the addition of reactant. The model of the ideal semibatch reactor assumes perfect mixing, as in the CSTR and batch reactor, but the reaction volume changes with time of operation due to the introduction of reactant and the method of processing is an unsteady-state one. A common utilization of the semibatch reactor is for reactions that are highly exothermic. The net rate of heat generation can be conveniently controlled by control of the rate of introduction of the second reactant.

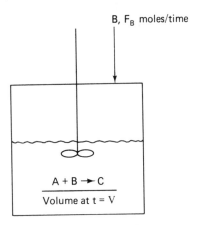

Figure 4.13 Semibatch reactor

Consider the process illustrated in Figure 4.13, where the second-order, irreversible reaction A + B $\longrightarrow$ C is carried out in a semibatch reactor. Now, since there is a large excess of A present, we may take the kinetics of the reaction to be pseudo-first-order in B, and since this is not a constant-volume process, let us write a molal balance on B.

$$\text{input} = F_B \qquad \text{moles B added/time (constant)}$$
$$\text{output} = (-r_B)V \qquad \text{moles B reacted/time}$$
$$\text{accumulation} = \frac{dN_B}{dt}$$

where V is the reaction mixture volume, $(-r_B)$ the rate of reaction per unit volume, and N_B the moles of B. Equating,

$$\frac{dN_B}{dt} = F_B - (-r_B)V \qquad (4\text{-}83)$$

If C_B is the concentration of B at any time, then

$$N_B = VC_B$$

and

$$\frac{dN_B}{dt} = V\frac{dC_B}{dt} + C_B\frac{dV}{dt}$$

Inserting this into equation (4-83), together with the first-order rate equation for B, we have

$$V\frac{dC_B}{dt} + C_B\frac{dV}{dt} = F_B - kC_BV \tag{4-84}$$

The volume of the reaction mixture (not the reactor volume) is included in equations (4-83) and (4-84), and this is given as a function of time by

$$V = V_A + F_B v_B t \tag{4-85}$$

where V_A is the volume of A initially present, v_B is the molal specific volume of B, and t is the time of the filling period, identical with the time of reaction. Substituting for V and dV/dt in equation (4-84),

$$(V_A + F_B v_B t)\frac{dC_B}{dt} + F_B v_B C_B = F_B - kC_B(V_A + F_B v_B t)$$

which on rearrangement becomes

$$\frac{dC_B}{dt} + C_B\left(\frac{1}{m+t} + k\right) = \frac{1/v_B}{m+t} \tag{4-86}$$

with $m = (V_A/F_B v_B)$. Equation (4-86) is linear and first-order, with initial condition that $C_B = 0$ at $t = 0$. The solution for C_B is

$$C_B = \frac{1 - e^{-kt}}{kv_B(m+t)} \tag{4-87}$$

In a similar manner we obtain for concentration of product, C_C:

$$C_C = \frac{t}{v_B(m+t)} + \frac{e^{-kt}-1}{kv_B(m+t)} \tag{4-88}$$

where $C_C = 0$ at $t = 0$.

For a reversible first-order reaction, $A + B \rightleftharpoons C$, under the same assumptions and conditions just treated, we obtain

$$C_B = \frac{1}{v_B(m+t)}\left[\frac{1 - e^{-(k_1+k_2)t}}{k_1+k_2} + k_2 t + \frac{k_2 e^{-(k_1+k_2)t}}{(k_1+k_2)^2} - \frac{k_2}{(k_1+k_2)^2}\right] \tag{4-89}$$

$$C_C = \frac{1}{v_B(m+t)}\left[\frac{k_1 t}{k_1+k_2} + \frac{k_1}{(k_1+k_2)^2}(e^{-(k_1+k_2)t} - 1)\right] \tag{4-90}$$

It is interesting to see, for a typical case, how well the assumption of pseudo-first-order kinetics based on $C_A \gg C_B$ stands up. Figure 4.14 gives some results in terms of the ratio of moles of A present to moles of B present for the reversible case. The results are specific to the particular parameters

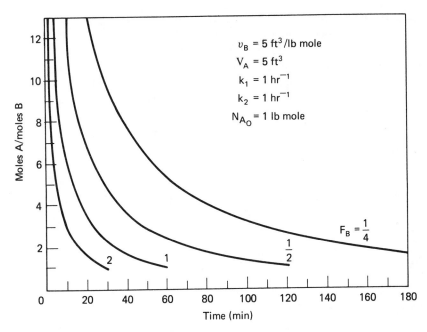

Figure 4.14 Molal reactants ratio for an example semibatch reactor

used, of course, but should illustrate that for times near the end of the filling period the assumption may break down. This depends ultimately on what is to be the total charge of B in comparison to the amount of A present initially.

For more complex kinetics than those illustrated, a general numerical procedure based on equation (4-83) is easily implemented. In finite-difference form,

$$\Delta N_B = [F_B - (-r_B)V]\,\Delta t \tag{4-91}$$

which relates the change in total moles of B over the time increment Δt. However, at the time of reaction, t, considered:

$$V = V_A + F_B v_B(t + \Delta t)$$

so that

$$\Delta N_B = \{F_B - (-r_B)[V_A + F_B v_B(t + \Delta t)]\}\,\Delta t \tag{4-92}$$

Initially, one picks an arbitrary (small) time increment Δt and calculates the corresponding ΔN_B. The rate of reaction, $(-r_B)$, which in all likelihood will be a function of C_B (and hence ΔN_B) is not known, so one must first guess its value. After ΔN_B has been calculated with the assumed value for $(-r_B)$, the concentration of B can be determined and a new value of $(-r_B)$ is then used to recalculate ΔN_B for the initial Δt, and the process repeated until there is no further change in $(-r_B)$ from one iteration to the next.

Calculations for successive time intervals follow the same procedure; $(-r_B)$ is known at the beginning of the interval, a value is assumed for the end of the interval, and ΔN_B corresponding is determined. The resulting C_B is used to check the assumed $(-r_B)$ at the end of the time interval. For a more precise calculation, one can average $(-r_B)$ over the time interval and use the check with C_B to determine a proper average rate over the interval considered.

d. Laminar flow reactors

In general, analytical solutions to the laminar flow reactor are not convenient to work with because of the awkward forms that arise from the concentration averaging of equation (4-55). For first-order irreversible kinetics, the design equation (4-54) becomes

$$\frac{1}{k} \int_{C_{A_0}}^{C_A} \frac{dC_A}{C_A} = -\frac{L}{u_0[1 - (r_p/R_0)^2]}$$

where u_0 is the central streamline velocity, $2V/\pi R_0^2$. Integrating the above we obtain the exit concentration as a function of radial position:

$$C_A = C_{A_0} \exp\left(-\frac{\lambda}{1 - U^2}\right) \tag{4-93}$$

where

$$\lambda = \frac{kL}{u_0} \qquad U = \frac{r_p}{R_0}$$

Substituting this result into equation (4-55) for the average exit concentration, we have

$$\langle C_A \rangle = C_{A_0} \frac{\int_0^1 U(1 - U^2)e^{-\lambda/(1-U^2)}\, dU}{\int_0^1 U(1 - U^2)\, dU}$$

$$= 4C_{A_0} \int_0^1 U(1 - U^2)e^{-\lambda/(1-U^2)}\, dU \tag{4-94}$$

The substitution $X = \lambda/(1 - U^2)$ allows partial integration of equation (4-94) to

$$\frac{\langle C_A \rangle}{C_{A_0}} = e^{-\lambda}(1 - \lambda) + \lambda^2 \int_\lambda^\infty \frac{e^{-X}\, dX}{X} \tag{4-95}$$

The integral in the second term is a form of the exponential integral, tabulated as $E_1(z)$ by Abramowitz and Stegun (M. Abramowitz and I. A. Stegun, *Handbook of Mathematical Functions*, Dover, New York, 1965) Further discussion of the ideal laminar flow reactor with a first-order reaction is given by Cleland and Wilhelm [F. A. Cleland and R. H. Wilhelm, *Amer. Inst. Chem. Eng. J.*, **2**, 489 (1956)].

A similar derivation for the second-order irreversible reaction $2A \rightarrow B$ has been given by Denbigh [K. G. Denbigh, *J. Appl. Chem.* (*London*), **1**, 227 (1951)]. The result corresponding to Equation (4-95) is

$$\frac{\langle C_A \rangle}{C_{A_0}} = 1 - 2\lambda'\left(1 - \lambda' \ln \frac{1 + \lambda'}{\lambda'}\right) \tag{4-96}$$

where $\lambda' = C_{A_0}kL/u_0$ and k is the appropriate second-order rate constant. The conversions for these two cases are illustrated in Figure 4.15.

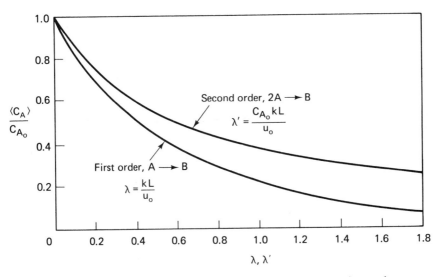

Figure 4.15 Conversion in a laminar flow reactor for first- and second-order irreversible reactions with no volume change

e. Some comparisons of conversion and selectivity in ideal reactors

Since the PFR and CSTR represent the extremes in mixing behavior, at least in terms of their residence time distributions, it is of interest to examine their relative performance characteristics for both single and multiple reactions. However, comparison solely on the basis of what the mathematical results of the reactor analysis tell us is often not the full story. Many factors associated with the operation, the nature of the process stream, the reaction conditions, and so on, are important in determination of the type of reactor to be used in a given situation. Stirred tank reactors are most often used for liquid-phase reactions, generally at low or moderate pressures. The nature of the economics of stirred tank reactors normally dictates that they will have relatively large volumes and correspondingly long average residence times, and, as we have mentioned before, the spatial uniformity of concentrations and temperature in this type of reactor provides an inherent ease of control for reactions involving large heat effects which is not possible to obtain with the tubular flow type. However, the heat-transfer rates in stirred tank reactors are generally lower than in tubular reactors, since their surface area/volume ratio is smaller, so that in spite of the inherent ease of control of CSTRs, tubular reactors are normally employed for reactions involving large heat-supply or heat-removal requirements. Gas-phase reactions are also not generally carried out in stirred tanks because of mechanical problems (stirrer seals effective at high temperature and pressure) and because of the difficulty of attaining well-mixed conditions in the gas phase. Most reactions employing heterogeneous catalysts are conducted in tubular reactors containing a fixed bed of the catalyst. These generally involve gas-phase process streams at moderate or high pressures and the reactions have substantial heat supply or removal requirements. Various types of gas/liquid and gas/liquid/solid reactions are conducted either in stirred reactors (slurry reactors, fluidized beds) or fixed beds (trickle beds), but these are specialized reactor types and we will not be concerned with their analysis at this point. An extensive discussion of the utilization of stirred tank reactors has been given in the papers by MacMullen and Weber and Piret and coworkers, previously cited, and by Denbigh and coworkers [K. G. Denbigh, *Trans. Faraday Soc.*, **40**, 352 (1944); **43**, 648 (1947); K. G. Denbigh, M. Hicks, and F. M. Page, *Trans. Faraday Soc.*, **44**, 479 (1948)].

For simple reactions in single reactors, the relative performance of the PFR and CSTR is most easily determined by comparing the residence times required by each to attain the same conversion level, which for a set feed rate then determines the reactor volume required. As an example, for a first-order irreversible reaction at constant density we have

$$t_{PFR} = -\frac{1}{k} \ell n\,(1-x)$$

and

$$\bar{t}_{CSTR} = \frac{1}{k}\frac{x}{1-x}$$

so that

$$\frac{t_{PFR}}{\bar{t}_{CSTR}} = \frac{(1-x)\,\ell n\,(1-x)}{x} \tag{4-97}$$

Corresponding results for zero-order and second-order reactions at constant density are

$$\text{zero order:} \quad \frac{t_{PFR}}{\bar{t}_{CSTR}} = 1 \tag{4-98}$$

$$\text{second order:} \quad \frac{t_{PFR}}{\bar{t}_{CSTR}} = 1-x \tag{4-99}$$
$$(2A \longrightarrow \text{products})$$

These results are shown in Figure 4.16, where it is seen that the PFR is more

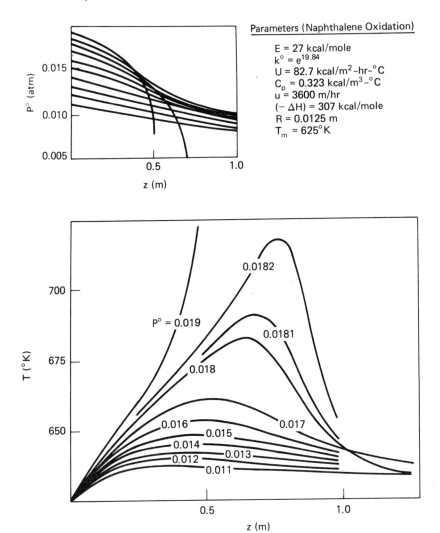

Parameters (Naphthalene Oxidation)

$E = 27$ kcal/mole
$k^\circ = e^{19.84}$
$U = 82.7$ kcal/m^2-hr-$^\circ$C
$C_p = 0.323$ kcal/m^3-$^\circ$C
$u = 3600$ m/hr
$(-\Delta H) = 307$ kcal/mole
$R = 0.0125$ m
$T_m = 625^\circ$K

Figure 4.20 Typical temperature and partial pressure profiles for non-isothermal reactor operation with a highly exothermic reaction (after R. J. Van Welsenaere and G. F. Froment, *Chem. Eng. Sci.*, **25**, 1503 (1970); with permission of Pergamon Press, Ltd., London)

is the extreme sensitivity of the locus and position of this hot spot to relatively small changes in inlet conditions. Such behavior, termed *parametric sensitivity*, will be discussed in more detail in chapter 6.

Adiabatic reactor problems may be handled by the same procedures and techniques developed for nonisothermal reactors by simply deleting the external heat transfer term from the energy balance. Also, Table 1.3 presented some analytical solutions for simple-order, irreversible adiabatic reactions in batch systems in terms of exponential integral functions. These solutions apply as well to the adiabatic plug-flow reactor if the system can be considered to be at constant density. In this event, time, t, in the relationships of Table 1.3 then becomes the PFR residence time. The analytical method has also been applied to the more practical problem of reactor operation at constant pressure. In Table 4.4 are given two examples for zero- and first-order reac-

TABLE 4.4

EXAMPLE ANALYTICAL SOLUTIONS FOR ADIABATIC PLUG-FLOW
REACTORS AT CONSTANT PRESSURE[a, b]

1. Zero order: $A \longrightarrow \nu P$

$$-r_A = k$$

$$\frac{\bar{V}}{F} = \frac{C_p}{k(-\Delta H)}\left[T \exp\left(\frac{E}{RT}\right) - T_0 \exp\left(\frac{E}{RT_0}\right)\right] - \frac{EC_p}{Rk(-\Delta H)}\cdot\left[E_i\left(\frac{E}{RT}\right) - E_i\left(\frac{E}{RT_0}\right)\right]$$

2. First order: $A \longrightarrow \nu$ products

$$\epsilon = \nu - 1 \qquad \delta = T_0 + \frac{(-\Delta H)n_{A_0}}{C_p}$$

$$-r_A = k_1 P\left(\frac{n_A}{n_t}\right)$$

$$\frac{\bar{V}}{F} = \frac{-\epsilon C_p}{k_1^\circ(-\Delta H)P}\left[T \exp\left(\frac{E}{RT}\right) - T_0 \exp\left(\frac{E}{RT_0}\right)\right]$$

$$+ \left[n_0 + \epsilon n_{A_0} + \frac{\epsilon EC_p}{R(-\Delta H)}\right](k_1^\circ P)^{-1}\left[E_i\left(\frac{E}{RT}\right) - E_i\left(\frac{E}{RT_0}\right)\right]$$

$$- \frac{(n_0 + \epsilon n_{A_0})\exp(E/R\delta)}{k_1^\circ P}[E_i(Z) - E_i(Z_0)]$$

where

$$Z = \frac{E}{R}\left(\frac{1}{T} - \frac{1}{\delta}\right) \qquad E_i(x) = \int_{-\infty}^{x}\frac{\exp x'\, dx'}{x'}$$

$$Z_0 = \frac{E}{R}\left(\frac{1}{T_0} - \frac{1}{\delta}\right)$$

$$n_t = n_0 + \epsilon x_A$$

[a] In the original reference the following dimensions are employed:

C_p = average heat capacity, Btu/mass-°F
F = feed rate, mass/time
$(-\Delta H)$ = heat of reaction, Btu/mole of A converted
k = zero order rate constant, moles A/ft^3-sec
k_1° = first order frequency factor, moles A/atm-ft^3-sec
n_{A_0} = initial moles of A per unit mass of feed
n_0 = total moles of feed per unit mass of feed
n_t = total moles of reacting system per unit mass of feed
x_A = moles of A converted per unit mass of feed
P = pressure, atm

[b] Any self-consistent set of dimensions may be employed. For example, the zero-order result may be expressed in F = moles fed/time if C_p is determined as a molar heat capacity.

where

$$E_1(x) = \int_x^\infty \frac{\exp(-x')}{x'} \, dx'.$$

A corresponding derivation for first-order reactions can be written

$$x = 1 - e^{-kt}$$

$$\bar{x} = \int_0^\infty (1 - e^{-kt}) \frac{e^{-t/\bar{t}}}{\bar{t}} \, dt \tag{5-4}$$

which gives

$$\bar{x} = \frac{k\bar{t}}{1 + k\bar{t}} \tag{5-5}$$

The result here is identical in form to the result obtained for perfect mixing, which is in agreement with the discussion early in chapter 4 showing that first-order reactions were unaffected by the extent of micromixing.

For half-order kinetics we may derive

$$x = 1 - \left(1 - \frac{kt}{2\sqrt{C_{A_0}}}\right)^2$$

for $0 < t < 2\sqrt{C_{A_0}}/k$. Then

$$\bar{x} = \int_0^{2\sqrt{C_{A_0}}/k} \left[1 - \left(1 - \frac{kt}{2\sqrt{C_{A_0}}}\right)\right] \frac{e^{-t/\bar{t}}}{\bar{t}} \, dt \tag{5-6}$$

which gives

$$\bar{x} = \frac{k\bar{t}}{(C_{A_0})^{1/2}}\left[1 - \frac{k\bar{t}}{2(C_{A_0})^{1/2}}(1 - e^{-2(C_{A_0})^{1/2}/k\bar{t}})\right] \tag{5-7}$$

Corresponding expressions for the case of complete mixing are given in Table 4.2. A comparison of typical conversion results for complete mixing and complete segregation for one-half order, first-order, and second-order kinetics is shown in Figure 5.2. It is seen that segregation *increases* average conversion for reaction orders > 1 and *decreases* conversion for reaction orders < 1, with no effect for first-order, as suggested previously. The difference in the curves for segregated and perfectly mixed cases indicates the magnitude of the effects due to micromixing. If some extent of micromixing exists in an actual situation, average conversions will lie between the two limits.

Direct use of the exit-age distribution in this manner, then, amounts to a segregated-flow model, no micromixing, which is able to predict for simple kinetics either the upper or lower bound of the effects of nonideality, depending on the order of the reaction.

The results previously obtained for conversions in a laminar flow reactor may also be obtained by use of the exit-age distribution function. We recall from chapter 4 the $E(t)$ for a laminar flow reactor:

$$E(t) = \frac{\bar{t}^2}{2t^3} \tag{4-53}$$

Curve: 1 − perfectly mixed, $n = \dfrac{1}{2}$

2 − segregated, $n = \dfrac{1}{2}$

3 − perfectly mixed or segregated, $n = 1$

4 − segregated, $n = 2$

5 − perfectly mixed, $n = 2$

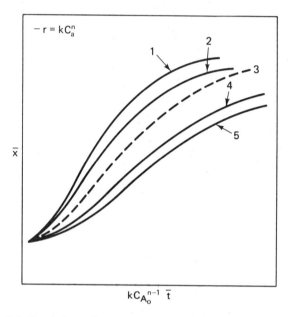

Figure 5.2 Comparison of conversions in a segregated-flow STR with a perfectly mixed STR for half-, first-, and second-order reactions

For a first-order reaction, then,

$$\bar{x} = \frac{\bar{t}^2}{2} \int_{\bar{t}/2}^{\infty} \frac{1 - e^{-kt}}{t^3} \, dt$$

and for a second-order case,

$$\bar{x} = \frac{kC_{A_0}\bar{t}^2}{2} \int_{\bar{t}/2}^{\infty} \frac{dt}{t^2(1 + ktC_{A_0})}$$

The demonstration that these forms reduce to the corresponding equations given previously [equations (4-95) and (4-96)] is left to the exercises.

In many applications in practice, such as shown in Figure 5.1, one will not have available analytical expressions for the exit-age distribution but rather concentration-response data from either pulse or step-input experimentation. In this case the integral of equation (5-1) is replaced by the corresponding summation and average conversion determined from

$$\bar{x} = \sum x(t)E(t)\,\Delta t \tag{5-8}$$

using the general techniques discussed in chapter 4.

b. Mixing-cell approximations

The thought on using mixing-cell models to approximate deviations from ideal behavior originates from studying the transient response of CSTR sequences. Let us run a tracer experiment, as described in chapter 4, for determining the RTD of a sequence of CSTR as shown in Figure 4.9. Initially, there is a steady flow of fluid throughout the sequence, and at a given time a step function of tracer is introduced. For any unit in the sequence, we have

$$\bar{V}_n \frac{dC_n}{dt} = v(C_{n-1} - C_n) \tag{5-9}$$

where $\bar{V}_n$ is the volume of the unit, C tracer concentration, and v the steady volumetric flow rate. This can be rearranged, for $\bar{V}_n$ all equal, to

$$\frac{dC_n}{dt} + \frac{C_n}{\bar{t}} = \frac{C_{n-1}}{\bar{t}} \tag{5-10}$$

where $\bar{t}$ is the nominal average residence time per unit, $\bar{V}/v$. For $n = 1$,

$$\frac{dC_1}{dt} + \frac{C_1}{\bar{t}} = \frac{C_0}{\bar{t}} \tag{5-11}$$

This is identical to equation (4-46), with the same limits of integration, with the result

$$C = C_0(1 - e^{-t/\bar{t}}) \tag{5-12}$$

Now, for $n = 2$,

$$\frac{dC_2}{dt} + \frac{C_2}{\bar{t}} = \frac{C_0(1 - e^{-t/\bar{t}})}{\bar{t}} \tag{5-13}$$

which has as its solution

$$C_2 = C_0\left[1 - e^{-2t/\bar{t}_t}\left(1 + \frac{2t}{\bar{t}_t}\right)\right] \tag{5-14}$$

where $\bar{t}_t$ is the mean residence time of the two unit sequence.

Following this procedure systematically throughout the sequence, one can obtain the general expression for unit n as

$$F_n(t) = F(\theta) = \frac{C_n}{C_0} = 1 - e^{-nt/\bar{t}_t}\left[\sum_{i=1}^{n-1} \frac{(nt/\bar{t}_t)^{i-1}}{(i-1)!}\right] \tag{5-15}$$

where $\bar{t}_t$ is the total residence time, $n\bar{t}$.

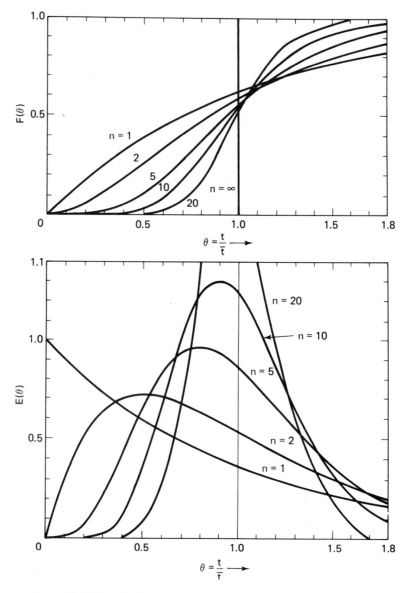

Figure 5.3 $F(\theta)$ and $E(\theta)$ as determined from mixing cells in series model (after A. R. Cooper and G. V. Jeffreys, *Chemical Kinetics and Reactor Design*, © 1971; reprinted by permission of Prentice-Hall, Inc., Englewood Cliffs, N.J.)

The exit-age distribution, $E_n(t)$, for the n-unit CSTR sequence is

$$E_n(t) = \frac{n^n}{\bar{t}_t} \left(\frac{t}{\bar{t}_t}\right)^{n-1} \frac{e^{-nt/\bar{t}_t}}{(n-1)!}$$ (5-16)

The shapes of $F(\theta)$ and $E(\theta)$ curves computed from these two expressions[1] are shown in Figure 5.3. It is clear that as n increases in the sequence the shapes of the responses assume the character of the intermediate cases illustrated in Figures 4.3 and 4.4. For very large n, the response becomes essentially that of the plug-flow situation.

When experimental data on $F(t)$ or $E(t)$ are available, one can obtain a model for the effects of nonideal behavior by fitting the response using n, the number of individual cells in the stirred tank sequence as an adjustable parameter. Then integral expressions conforming to equation (5-1) may be set up directly in terms of $E(t)$ from equation (5-16) for computing conversions. For example, consider the conversion in a second-order reaction. From equation (5-1) we may write

$$\bar{x} = \int_0^\infty \frac{kC_{A_0}t}{1 + kC_{A_0}t} \cdot \frac{n^n}{\bar{t}_t}\left(\frac{t}{\bar{t}_t}\right)^{n-1}\frac{e^{-nt/\bar{t}_t}}{(n-1)!}\,dt$$

This may be rearranged to

$$\bar{x} = \frac{n^n}{(\bar{t}_t)^n(n-1)!}\int_0^\infty \frac{t^n e^{-nt/\bar{t}_t}\,dt}{(1/kC_{A_0}) + t}$$

and on integration,

$$\bar{x} = \frac{n^n}{(\bar{t}_t)^n(n-1)!}\left[(-1)^{n-1}\left(\frac{1}{kC_{A_0}}\right)\exp\left(\frac{nt}{kC_{A_0}\bar{t}_t}\right)E_i\left(-\frac{nt}{kC_{A_0}\bar{t}_t}\right)\right.$$
$$\left.+ \sum_{i=1}^{n}(i-1)!\left(-\frac{1}{kC_{A_0}}\right)^{n-i}\left(\frac{nt}{\bar{t}_t}\right)^{-i}\right]$$

As is seen from this example, for most types of kinetics the integrals are a bit messy for analytic evaluation, but numerical evaluation is still quite simple. Again, we point out that this amounts to a segregated-flow model which has the capability of predicting upper or lower bounds on conversion depending on whether kinetics, assuming simple orders, are greater or lesser than first-order, respectively.

A simple modification of the mixing-cell sequence has been proposed which would provide for the incorporation of micromixing effects into the model. In this modification we view that there is some circulation between individual cells, as shown in Figure 5.4. This flow in the reverse direction, or backmixing, is equivalent to a micromixing effect in the reactor model. We will let f be the fraction of fluid involved in recirculation expressed in terms of the net volumetric flow v. Then, for the first cell in the series, our tracer

[1]Remember that $E(\theta) = \bar{t}_t E(t)$.

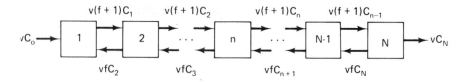

Individual cell volume = $\bar{V}$
Net volumetric flow rate = v
Frictional recirculation between stages = f

Figure 5.4 Mixing-cell sequence with micromixing effects

experiment material balance is

$$\bar{V}\frac{dC_1}{dt} = vC_0 - v(f+1)C_1 + vfC_2 \qquad (5\text{-}17)$$

and for the nth cell,

$$\bar{V}\frac{dC_n}{dt} = v(1+f)(C_{n-1} - C_n) + vf(C_{n+1} - C_n) \qquad (5\text{-}18)$$

It is apparent immediately that generation of analytical $F(t)$ or $E(t)$ responses is not going to be a fruitful exercise with this model, since the concentration of tracer downstream, C_{n+1}, appears in the equation for determination of C_n. However, there is an interesting relationship between these CSTR sequence mixing models, both with and without backmixing, and the dispersion models to be discussed below, so both versions have been presented here.

c. Dispersion models

We have already called attention to the fact that the spreading observed in pulse-response experimentation is reminiscent of diffusional phenomena, so it is reasonable to expect that a diffusional "correction term" appended to the plug-flow model would also afford a means for modeling nonideal flows. Consider the one-dimensional dispersion model as illustrated in Figure 5.5. For the mass balance around the differential length dz:

in: convection: uCA

             dispersion: $-D\dfrac{dC}{dz} \cdot A$

out: convection: $u\left(C + \dfrac{dC}{dz} \cdot dz\right)A$

             dispersion: $-D\left[\dfrac{dC}{dz} + \dfrac{d}{dz}\left(\dfrac{dC}{dz}\right)dz\right]A$

accumulation: $\dfrac{dC}{dt} \cdot A\,dz$

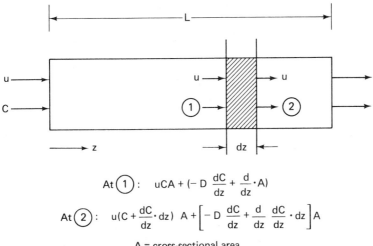

$$\text{At } \textcircled{1}: \quad uCA + (-D\,\frac{dC}{dz} + \frac{d}{dz}\cdot A)$$

$$\text{At } \textcircled{2}: \quad u(C + \frac{dC}{dz}\cdot dz)\ A + \left[-D\,\frac{dC}{dz} + \frac{d}{dz}\,\frac{dC}{dz}\cdot dz\right]A$$

A = cross-sectional area

Figure 5.5 One-dimensional dispersion model

Equating the difference in input/output terms to accumulation in the volume element gives

$$\frac{dC}{dt}\cdot dz = -u\frac{dC}{dz}\cdot dz + D\frac{d}{dz}\left(\frac{dC}{dz}\right)dz \tag{5-19}$$

which on rearrangement becomes

$$D\frac{d^2C}{dz^2} - u\frac{dC}{dz} = \frac{dC}{dt} \tag{5-20}$$

Here D, the axial diffusion (or dispersion) coefficient, is the parameter employed to describe the deviations from ideal flow. If u is taken to be constant in the radial direction, the rightmost terms in equation (5-20) constitute the plug-flow mixing model [equation (4-37)] and $D(d^2C/dz^2)$ a Fickian form of diffusional correction term.

Now let us see what this model gives us in terms of $F(t)$ or $E(t)$ responses. To solve this equation (which incidentally is no longer of the form of an initial-value problem but is a boundary-value problem) it is convenient to make the following change of variables. Let v represent the position of the moving interface represented by all elements of fluid introduced into the reactor at some given time. In terms of the length variable, this transformation is

$$v = z - ut \tag{5-21}$$

Substituting for z in terms of v in equation (5-20), we obtain

$$D\frac{d^2C}{dv^2} = \frac{dC}{dt} \tag{5-22}$$

The boundary conditions for a step function in inlet concentration of tracer are[1]

$$
\begin{aligned}
C &= 0 && (v > 0, t = 0) \\
C &= 1 && (v < 0, t = 0) \\
C &= 0 && (v = \infty, t > 0) \\
C &= 1 && (v = -\infty, t > 0)
\end{aligned}
\tag{5-23}
$$

The corresponding solution is

$$
C = \frac{1}{2}\left[1 - \mathrm{erf}\left(\frac{v}{2\sqrt{Dt}}\right)\right]
\tag{5-24}
$$

where

$$
\mathrm{erf}\,(y) = \frac{2}{\sqrt{\pi}} \int_0^y \exp\,(-v^2)\, dv
$$

and is, again, a tabulated function. The $F(t)$ response computed from the one-dimensional dispersion model is, then,

$$
F(t) = \frac{1}{2} - \frac{1}{2}\,\mathrm{erf}\left(\frac{L - ut}{2\sqrt{Dt}}\right)
\tag{5-25}
$$

If we rewrite this result in terms of the residence time, $t_R = L/u$:

$$
F(t) = F(\theta) = \frac{1}{2} - \frac{1}{2}\,\mathrm{erf}\left[\frac{1 - t/t_R}{2\sqrt{(t/t_R)(D/Lu)}}\right]
\tag{5-26}
$$

The dimensionless grouping, (Lu/D), appearing in equation (5-26) has commonly been used to represent the magnitude of dispersion effects, since it represents the ratio of characteristic time constants for convective and dispersive effects. It is termed the *axial Peclet number*, N_{Pe}. Now equations (5-25) or (5-26) can be fit to experimental $F(t)$ data to determine the value of D or N_{Pe} applicable to a given situation. This can be done either with a least-squares procedure, or at least as a first approximation, by determining the midpoint slope of the $F(t)$ data. The latter method derives from the fact that

$$
\left[\frac{d[F(t)]}{d(t/t_R)}\right]_{t/t_R=1} = \frac{1}{2}\sqrt{\frac{Lu}{\pi D}} = \frac{1}{2}\sqrt{\frac{N_{Pe}}{\pi}}
\tag{5-27}
$$

so numerical estimation of $\Delta F(t)/\Delta(t/t_R)$ at the point $(t/t_R) = 1$ yields directly a value for N_{Pe}.

Solution of the dispersion equation for the impulse response, using a slightly different set of boundary conditions for equation (5-20), gives

$$
E(\theta) = \frac{\exp\{-[1 - (t/t_R)]^2/4(D/uL)(t/t_R)\}}{2\sqrt{\pi(D/uL)(t/t_R)}}
\tag{5-28}
$$

---

[1]Considerable energy has been expended in the literature debating the proper form for these boundary conditions. The question will be addressed in more detail in a later section. "Now we see through a glass, darkly" (1 Corinthians XIII, 12).

In the case where dispersion effects are small, the expression for $E(\theta)$ given above can be approximated by a Gaussian form:

$$E(\theta) = \frac{1}{2\sqrt{\pi(D/uL)}} \exp\left[-\frac{(1 - t/t_R)^2}{4(D/uL)}\right] \tag{5-29}$$

Examples of $F(\theta)$ and $E(\theta)$ curves computed from equations (5-26) and (5-28) are given in Figures 5.6a and b. It is seen that their shapes are quite similar to those obtained from the mixing cell in series approach, but in this case the Peclet number rather than the number of mixing cells is the parameter determining the shape.

Now, the way we have presented the one-dimensional dispersion model here so far has been as a modification of the plug-flow model. Hence, u is treated as uniform across the tubular cross section. In fact, the general form of the model can be applied to numerous instances when this is not so, in which case the dispersion coefficient D becomes a parameter describing the net effect of a number of different phenomena. This is nicely illustrated by the early work of Taylor [G. I. Taylor, *Proc. Roy. Soc.* (*London*), **A219**, 186 (1953); **A223**, 446 (1954); **A224**, 473 (1954)] on the combined contributions of the velocity profile and molecular diffusion to the residence-time distribution for laminar flow in a tube. The situation considered is illustrated in

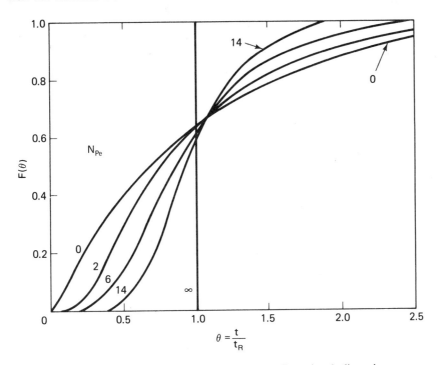

Figure 5.6a $F(\theta)$ as determined from the one-dimensional dispersion model

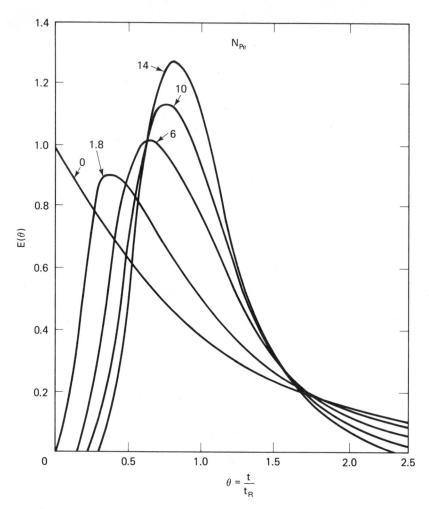

Figure 5.6b $E(\theta)$ as determined from the one-dimensional dispersion model

Figure 5.7. For moderate flow velocities the dispersion of a tracer in laminar flow will occur by axial and radial diffusion from the flow front and, in the absence of eddy motion, this will be via a molecular diffusion mechanism. However, the net contribution of diffusion in the axial direction can be taken as small in comparison to the contribution of the flow velocity profile. This leaves us with a two-dimensional problem, diffusion in the radial direction and convection in the longitudinal direction. Following the procedures outlined for equation (5-20), we may derive the following equation for this case:

$$\frac{\partial C}{\partial t} = D_M \left(\frac{\partial^2 C}{\partial r^2} + \frac{1}{r} \frac{\partial C}{\partial r} \right) - u(r) \frac{\partial C}{\partial z} \qquad (5\text{-}30)$$

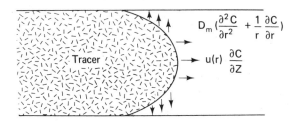

Figure 5.7 Convective and molecular diffusion contributions to dispersion in laminar flow

where D_M is a molecular diffusivity for the tracer and $u(r)$ represents the laminar flow velocity profile. Now equation (5-30) is a nonsteady-state partial differential equation, the solution of which under the best of circumstances is going to be a demanding task. As engineers, we often look for rational approximations or equivalent representations in models which can save us work.[1] This is precisely what Taylor was able to do. He showed that an exactly equivalent representation of equation (5-30) is given by

$$\frac{\partial C_m}{\partial t} = D_e \frac{\partial^2 C_m}{\partial z^2} - \bar{u} \frac{\partial C_m}{\partial z} \tag{5-31}$$

where C_m is a mean radial concentration, $\bar{u}$ the average velocity independent of r, and D_e an effective dispersion coefficient. In terms of the parameters of equation (5-30) D_e is given by

$$D_e = \frac{R^2 \bar{u}^2}{48 D_m} \tag{5-32}$$

where R is the tubular radius. Of course, this representation is valid only under the conditions stated, with molecular diffusion in the axial direction being negligible. Taylor further showed that, for this condition to exist, it was necessary that

$$\frac{4L}{R} \gg \frac{\bar{u}R}{D_m} \gg 6.9 \tag{5-33}$$

Note in the inequality of (5-33) that the quantity $(\bar{u}R/D_m)$ can be considered a radial Peclet number, defined analogously to the longitudinal quantity (uL/D) used previously. The right-hand bound of 6.9 has been criticized by Ananthakrishnan et al. [V. Ananthakrishnan, W. N. Gill, and A. I. Barduhn, *Amer. Inst. Chem. Eng. J.*, **11**, 1063 (1965)] as not being sufficiently conservative. They propose that $(\bar{u}R/D_m) > 50$ should be used as a criterion for the use of the one-dimensional approximation.

Subsequently Aris [R. Aris, *Proc. Roy. Soc. (London)*, A235, 67 (1956)] showed that the effects of molecular diffusion in the axial direction could be

---

[1] If we are able to come up with good approximations or equivalent representations, we are good engineers. If the substitutes are somewhat shoddy, but we use them anyway, then we are lazy. "Drowsiness shall clothe a man with rags" (Proverbs, XXIII, 21).

included in the one-dimensional representation of equation (5-31) and that, in fact, the correction was a simple additive term:

$$D_e = D_m + \frac{R^2 \bar{u}^2}{48 D_m} \tag{5-34}$$

Aris also generalized the analysis to include all types of velocity distributions in flow vessels of any geometry.

These one-dimensional dispersion models for laminar flow are of importance for two reasons: (1) much industrial processing involving polymers, oils, colloids, pastes, and the like involves laminar flow and in many cases non-Newtonian fluids; and (2) the fact that a true two-dimensional problem can be reduced to an equivalent one-dimensional problem—although admittedly for a rather special set of circumstances—lends credibility to the extensive use of the one-dimensional dispersion model reported in the literature and reflected in our later discussion of chemical reactor modeling. Analytical solutions and one-dimensional approximations are reviewed and discussed in detail by Wen and Fan (C. Y. Wen and L. T. Fan, *Models for Flow Systems and Chemical Reactors*, Marcel Dekker, New York, 1975) for the following cases involving laminar flow:

1. Molecular diffusion negligible compared to the contribution of the velocity profile (for Reynolds numbers in the upper range of the laminar flow regime).
2. Both radial and axial molecular diffusion effects comparable to the contribution of the velocity profile (midrange of Reynolds number).

The analysis presented here pertains to case 2 above; derivation of the $F(t)$ relationship for laminar flow reactors given in chapter 4 pertains to case 1, a segregated-flow model.

d. Comparison of mixing-cell and dispersion models

In chapter 4 it was pointed out that the performance of a CSTR sequence approached that of a PFR of equivalent total residence time as the number of units in the sequence approached infinity. This result is also indicated by the $F(\theta)$ and $E(\theta)$ curves computed from the mixing-cell model reported in Figure 5.3. Now since the plug-flow model represents one limit of the dispersion model, that when $D \to 0$, it is reasonable to assume that there is an interrelationship between mixing cell and dispersion models which can be elucidated in the more general case of finite values for D. Let us start with the unsteady mass balance of the one-dimensional dispersion model:

$$D \frac{d^2 C}{dz^2} - u \frac{dC}{dz} = \frac{dC}{dt} \tag{5-20}$$

If we use a central finite-difference approximation to represent the spatial derivatives in equation (5-20), the following result is obtained for point n:

$$D\frac{C_{n-1} - 2C_n + C_{n+1}}{(\Delta z)^2} - \frac{u}{2\,\Delta z}(C_{n+1} - C_{n-1}) = \frac{dC_n}{dt} \qquad (5\text{-}35)$$

On the other hand, the mass balance for the mixing cells in series model is

$$v(C_{n-1} - C_n) = \bar{V}_n\frac{dC_n}{dt} \qquad (5\text{-}9)$$

Now let us view the dispersion model in terms of a series of mixing cells, of as-yet-unspecified dimension, and determine the conditions required for equivalence between the two. Thus, we may rewrite equation (5-9) as

$$\frac{dC_n}{dt} = \frac{\bar{u}C_{n-1} - \bar{u}C_n}{2\,\Delta z} \qquad (5\text{-}36)$$

where Δz represents some length over which perfect mixing occurs, and $\bar{u}$ is $(1/\bar{t})$. If we add and subtract the term $\Delta z\bar{u}\,C_{n+1}/2\,(\Delta z)^2$ to equation (5-36)

$$\frac{dC_n}{dt} = \frac{\bar{u}\,\Delta z}{(2\Delta z)^2}(C_{n+1} - 2C_n + C_{n-1}) - \frac{\Delta z\bar{u}}{2(\Delta z)^2}(C_{n+1} - C_{n-1}) \qquad (5\text{-}37)$$

Term-by-term comparison of equations (5-37) and (5-35) discloses the following requirement for equivalence of the two models:

$$\frac{\bar{u}\,\Delta z}{2} = D, \ (C_n \approx C_{n=1}) \qquad (5\text{-}38)$$

Thus, the axial Peclet number correct for defining the equivalent length of a perfect mixing cell is 2. This, in turn, provides the relationship between the number of mixing-cell parameters of the mixing-cell model and the corresponding axial dispersion coefficient of the dispersion model.

A similar comparison may be made between the dispersion model and the forward/reverse flow mixing-cell model (Figure 5.4). This perhaps is a physically more meaningful comparison, even though we were unable to use the model to derive analytical expressions for $F(t)$ and $E(t)$, because the forward/backward communication provided for is more akin to the physical nature of the diffusion process. In this case the expression corresponding to equation (5-37) is

$$\frac{dC_n}{dt} = \frac{f\bar{u}}{\Delta z}(C_{n-1} - 2C_n + C_{n+1}) - \frac{\bar{u}}{2\,\Delta z}(C_{n+1} - C_n) \qquad (5\text{-}39)$$

The equivalence condition is

$$\frac{f\bar{u}}{\Delta z} \approx \frac{D}{(\Delta z)^2} = \frac{L/\Delta z}{N_{\text{Pe}}} \qquad (5\text{-}40)$$

Here it is clear that the parameters Δz and f are interrelated through the equivalence condition.

The fact that such equivalencies between the mixing-cell and dispersion models exist makes the task of correlating information on nonideal flows somewhat more simple. In the following we shall refer to the Peclet number as the measure of the influence of nonideal flow with the understanding that modeling can be accomplished interchangably with either dispersion or mixing-cell approaches.

e. Evaluation of the Peclet number as a parameter in nonideal flow modeling

Over approximately the last 25 years a considerable body of experimental information has been built up concerning values of the Peclet number, or the corresponding dispersion coefficient, for various types of flow situations. These include laminar and turbulent flow of both gases and liquids in empty tubes, fixed and fluidized beds, and extraction columns. The bulk of available data deals with axial dispersion, but there have also been a number of studies of radial dispersion, primarily for liquids and gases in fixed beds. These results have been summarized in great detail in the monograph of Wen and Fan.

For laminar flow, the analysis of Taylor and Aris led to the result

$$D_e = D_m + \frac{R^2 \bar{u}^2}{48 D_m} \tag{5-34}$$

which can be rewritten in terms of the Reynolds, Schmidt, and Peclet numbers as

$$\frac{1}{N_{Pe}} = \frac{1}{N_{Re} N_{Sc}} + \frac{N_{Re} N_{Sc}}{192} \tag{5-41}$$

where $N_{Pe} = \bar{u} d / D_e$, $N_{Re} = d \bar{u} \rho / \mu$, and $N_{Sc} = \mu / \rho D_M$. Note that the Peclet number here has been written on the basis of tube diameter, d, rather than radius to make it consistent with the normal definition of the Reynolds number. Available laminar flow data in empty tubes, indicated by the cross-hatched region, are plotted in comparison with equation (5-41) in Figure 5.8a for the range $1 < N_{Re} < 2000$ and for $0.23 < N_{Sc} < 1000$; it is seen that the agreement is excellent. For turbulent flow in empty tubes, an approximate relationship between N_{Pe} and N_{Re} derived from the original arguments of Taylor is

$$\frac{1}{N_{Pe}} = \frac{1}{(N_{Re})^{0.125}} \tag{5-42}$$

An empirical correlation based on this expression, accounting for the transition regime as well, is given as

$$\frac{1}{N_{Pe}} = \frac{3 \times 10^7}{(N_{Re})^{2.1}} + \frac{1.35}{(N_{Re})^{0.125}} \tag{5-43}$$

The fit to experiment afforded by equation (5-43) together with the limiting value of $(N_{Re})^{-0.125}$ is given in Figure 5-8b. It is seen that N_{Pe} is not a function of N_{Sc} in the region of well-developed turbulence.

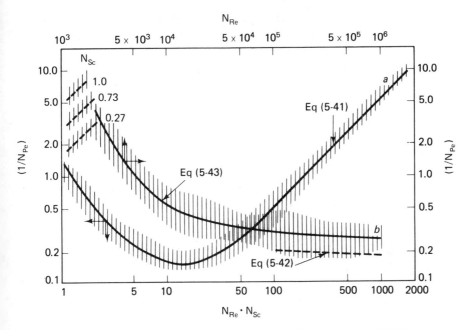

Figure 5.8 (a) Correlation of axial dispersion coefficient for flow of fluids through pipes in laminar flow region ($N_{Re} < 2000$); **(b)** Correlation of axial dispersion coefficient for flow of fluids through pipes for $N_{Re} > 2000$ (shaded area represents approximate range of experimental data)

Flow of liquids or gases through fixed beds is very important in chemical reactor engineering, since many commercially important processes involve reactors that contain beds of catalyst used to promote a desired reaction. The axial dispersion model has been extensively used to model these flows, even though two phases, fluid and solid, are present. Such a pseudo-homogeneous model assumes the same form we have described in the preceding section if the Peclet number is based on particle diameter and the interstitial fluid velocity is used. In this case

$$N_{Pe} = \frac{\bar{u} d_p}{D_e} \tag{5-44}$$

where $\bar{u} = u_0/\epsilon$, $\bar{u}$ is the interstitial velocity, u_0 the superficial velocity, and ϵ the bed porosity. Similar types of models have also been applied to the description of fluidized beds, although with somewhat less success, owing to the far more complex mixing problems associated with these beds. Chung and Wen [S. F. Chung and C. Y. Wen, *Amer. Inst. Chem. Eng. J.*, **14**, 857 (1968)] have given a correlation for liquids based on the pseudo-homogeneous model of both fixed and fluidized bed data, shown in Figure 5.9a. Though the correlation is based on both fixed and fluidized bed data, much of the scatter arises from the latter and we shall regard the correlation primarily of use for fixed beds. Further description of extension of the correlation to fluid beds is given by Chung and Wen. The correlation equation corresponding to

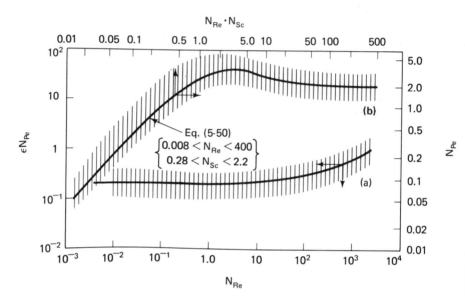

Figure 5.9 (a) Correlation of axial dispersion coefficients in liquid phase fixed and fluidized beds; **(b)** Correlation of axial dispersion coefficient for gases flowing through fixed beds

Figure 5.9a is

$$\epsilon N_{Pe} = 0.2 + 0.011(N_{Re})^{0.48} \tag{5-45}$$

Note that N_{Re} in the figure and in equation (5-45) is defined on the basis of superficial velocity, $N_{Re} = d_p u_0 \rho / \mu$. Owing to the low values of the molecular diffusion coefficient in liquids, the Schmidt number is not an important contributor to N_{Pe} in this case.

For gaseous flows in fixed beds, we can develop the approximate form of correlation to expect based on the following arguments. First, owing to the larger magnitude of diffusion coefficients in the gas phase, it is reasonable to expect that this will be a contributing factor at low N_{Re}. This can be expressed in terms of the molecular diffusion coefficient via a tortuosity factor, τ, which accounts in some gross sense for the reduction of diffusivity occasioned by the nondirect flow paths characteristic of the fixed bed. Thus, at very low flow velocities through a bed of porosity ϵ, we may expect

$$D_e = \frac{D_m \epsilon}{\tau} \tag{5-46}$$

On the other hand, it has been shown that at very large values for N_{Re} the following obtains:

$$N_{Pe} = \frac{\bar{u}d_p}{D_e} = \frac{2}{\eta} \tag{5-47}$$

where the distance between successive layers of particles is given by ηd_p and $0.8 \lesssim \eta \lesssim 1$ for various types of packing geometry. For our present purposes, if we take $\eta = 1$, then at high N_{Re},

$$D_e = \frac{\bar{u}d_p}{2} \tag{5-47a}$$

Combining low- and high-velocity contributions to D_e additively gives

$$D_e = \frac{D_m \epsilon}{\tau} + \frac{\bar{u}d_p}{2} \tag{5-48}$$

which can be rewritten in terms of the corresponding dimensionless numbers as

$$\frac{1}{N_{Pe}} = \frac{\epsilon(\epsilon/\tau)}{N_{Re}N_{Sc}} + \frac{1}{2} \tag{5-49}$$

The experimental data shown in Figure 5.9b exhibit a slight maximum for values of $N_{Re}N_{Sc}$ in the range 1 to 10, a feature not to be expected from the form of equation (5-49). Aside from this, however, the limiting behavior expected is observed and an empirical correlation based on equation (5-49) is

$$\frac{1}{N_{Pe}} = \frac{0.3}{N_{Sc}N_{Re}} + \frac{0.5}{1 + 3.8(N_{Re}N_{Sc})^{-1}} \tag{5-50}$$

where N_{Re} is again based on the superficial fluid velocity.

Thus far the emphasis has been on one-dimensional dispersion models, for reasons we have discussed previously. However, in some cases, particularly in the modeling of large-scale nonisothermal reactors, the one-dimensional approach is not sufficient. Indeed, the evaluation of conditions corresponding to the adequacy of a given model is a delicate art, as has been shown by Froment [G. F. Froment, *Ind. Eng. Chem.*, **59**, 18 (1967)]. In nonisothermal operation, the existence of significant radial temperature gradients will induce corresponding radial concentration gradients and we must resort to a two-dimensional dispersion representation. In terms of a mixing model only (no reaction), this is

$$\frac{\partial C}{\partial t} = D_r \left(\frac{\partial^2 C}{\partial r^2} + \frac{1}{r} \frac{\partial C}{\partial r} \right) + D_e \frac{\partial^2 C}{\partial z^2} - \bar{u} \frac{\partial C}{\partial z} \tag{5-51}$$

Values of the radial dispersion coefficient, or the corresponding radial Peclet numbers, $(\bar{u}d_p/D_r)$, in packed beds have been determined for both liquids and gases by a number of workers, and they are not the same as values in the axial direction. These results are shown in Figure 5.10a and b for liquids and gases, respectively. The corresponding empirical equations fitting these data are

$$N_{Pe_r} = \frac{17.5}{(N_{Re})^{0.75}} + 11.4 \qquad \text{(liquids)} \qquad (5\text{-}52)$$

$$\frac{1}{N_{Pe_r}} = \frac{0.4}{(N_{Pe}N_{Sc})^{0.8}} + \frac{0.09}{1 + (10/N_{Re}N_{Sc})} \qquad \text{(gases)} \qquad (5\text{-}53)$$

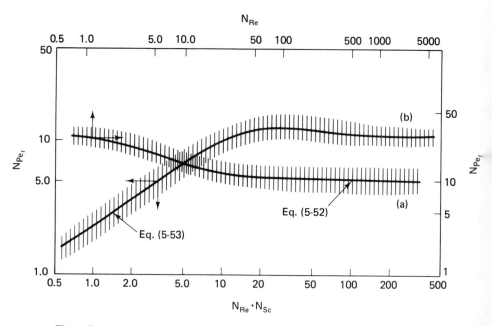

Figure 5.10 (a) Correlation of radial dispersion coefficient for liquids in fixed beds; (b) Correlation of radial dispersion coefficient for gases in fixed beds

As with the results for axial N_{Pe}, the value of N_{Sc} is important only for dispersion in gaseous systems.

A fact that is convenient to keep in mind is the limiting value of N_{Pe} for axial and radial directions in fixed beds, particularly for gases. At the higher values of N_{Re}, which in general are more typical of industrial processing conditions, N_{Pe} axial approaches 2, while N_{Pe} radial is approximately 10. It is easy to show, at least for the one-dimensional dispersion model, that this represents only a small deviation from plug-flow results. For example, the axial Peclet number appearing in equation (5-26) for the $F(t)$ of the one-dimensional dispersion model in an empty tube is defined on the basis of total length L. The axial Peclet numbers we have just been considering for flow in packed beds are defined on the basis of some representative particle diameter, d_p. In most applications of fixed-bed reactors the bed consists of many hundreds or even thousands of layers of catalyst particles; if there are, say, j layers of particles, then the Peclet number based on total length is $2j$ in the limit of higher N_{Re}. This would approach the plug-flow response for sufficiently large j (Figure 5.6a).

In spite of this consideration, however, we will not dismiss these deviations from ideality as being unimportant in practical application, since it will be seen later that even if conversion in a given reactor is little affected by dispersion, selectivity may be. Further, in nonisothermal reactors analogous mechanisms for the dispersion of energy significantly affect the shape of temperature profiles within the reactor. Since temperature is the single most significant kinetic variable, this will have a profound effect on reactor performance.

5.3   Combined Models for Macroscopic Flow Phenomena

The mixing-cell approximations or dispersion models have the ability to reproduce $F(t)$ or $E(t)$ responses intermediate between the limiting PFR and CSTR models (Figure 4.4a and b). Yet, they are clearly incapable of dealing with some of the very anomalous behavior induced by problems such as channeling or bypassing shown in Figure 5.1, because they assume some type of symmetry along the direction of flow. We can remove this restriction in modeling if we use combined models consisting of assemblies of individual (and differing) mixing models in various sequences. The original proposal was that of Cholette and Cloutier [A. Cholette and L. Cloutier, *Can. J. Chem. Eng.*, **37**, 105 (1959)], who envisioned three basic contributions to a given flow: (1) plug flow, (2) short-circuiting, and (3) complete mixing. We may conveniently add to this dead volume as well.

Now by assembling various combinations of plug flow, complete mixing, short-circuiting, and dead volume we may simulate a very large number of different types of macroscopic flow phenomena. Derivation of the appropriate response functions for such models is not difficult, since they essentially consist of assembling components from the types of analysis we have carried out previously. In Table 5.1 are given several examples of combined flow models together with their $F(t)$ responses. Consider the CSTR with short-circuiting and dead volume, number 5 in the table. The contribution of the perfect mixing section to $F(t)$ would be that of a CSTR of effective volume $f\bar{V}$, where f is the fraction of total volume perfectly mixed, multiplied by the fraction of total flow passing through the mixing circuit, v_1/v. Thus,

$$F(t)(\text{mixing} + \text{dead volume}) = \frac{v_1}{v}\left[1 - \exp\left(-\frac{v_1}{v}\cdot\frac{t}{f\bar{t}}\right)\right]$$

TABLE 5.1
SOME TYPICAL COMBINED FLOW MODELS

1. Plug flow with dead space:

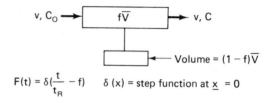

$$F(t) = \delta(\frac{t}{t_R} - f) \qquad \delta(x) = \text{step function at } \underline{x} = 0$$

2. CSTR with dead space:

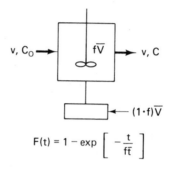

$$F(t) = 1 - \exp\left[-\frac{t}{f\bar{t}} \right]$$

3. CSTR with short-circuiting:

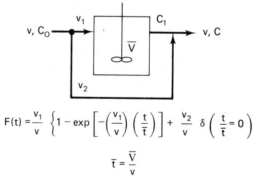

$$F(t) = \frac{v_1}{v} \left\{ 1 - \exp\left[-\left(\frac{v_1}{v}\right)\left(\frac{t}{\bar{t}}\right) \right] + \frac{v_2}{v} \delta\left(\frac{t}{\bar{t}} = 0\right) \right.$$

$$\bar{t} = \frac{\bar{V}}{v}$$

278

TABLE 5.1 (cont.)

4.   PFR-CSTR sequence:

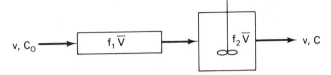

$f_1 + f_2 = 1$; $\overline{V}$ = combined PFR-CSTR volume

$$t_R = \frac{f_1 \overline{V}}{v}, \quad \overline{t} = \frac{f_2 \overline{V}}{v}$$

$$F(t) = 1 - \exp\left[\left(\frac{f_1}{f_2}\right) - \frac{t}{f_2(t_R + \overline{t})}\right] \; \delta\left(\frac{t}{t_R + \overline{t}} - f_1\right)$$

This expression applies for either PFR-CSTR or CSTR-PFR sequence

5.   CSTR with short-circuiting and dead volume:

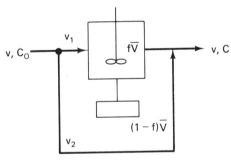

$$F(t) = \frac{v_1}{v}\left[1 - \exp\left(-\frac{v_1}{fv} \cdot \frac{t}{\overline{t}}\right)\right] + \frac{v_2}{v} \; \delta\left(\frac{t}{\overline{t}} = 0\right)$$

$$\overline{t} = \frac{\overline{V}}{v}$$

where the residence time $\overline{t}$ is based on the total reactor volume, $\overline{V}$. The contribution of the short circuit is direct transfer of the volume fraction of feed (v_2/v) to the outlet, effective at time zero. Thus,

$$F(t)(\text{short circuit}) = \frac{v_2}{v}$$

and the net residence time distribution for the combined model is

$$F(t) = \frac{v_1}{v}\left[1 - \exp\left(-\frac{v_1}{fv} \cdot \frac{t}{\bar{t}}\right)\right] + \frac{v_2}{v} \qquad (5\text{-}54)$$

Most of the more simple of these models involve exponential functions for their $F(t)$ responses as shown in the table, so tests of their possible application in interpretation of $F(t)$ data are conveniently carried out via log-linear plots of the response data. If straight lines are obtained, the model parameters are readily evaluated.

Obviously many such combinations can be assembled and the intelligent application of a given combined model to interpretation of nonideal flows in a given reactor must rely on additional information concerning geometric properties such as stirrer placement, feed inlet and product withdrawal placement, corners or internal structural elements leading to dead volumes, and so on. A certain amount of intuition is also a useful commodity in such modeling.[1] It is apparent from the models given in Table 5.1, for example, that none of them will be able to reproduce the $F(t)$ or $C(t)$ response typical of channeling shown in Figure 5.1. Examination of this response indicates a bimodal $C(t)$, each peak of which is generally characteristic of a mixing-cell response characterized by different nominal residence times. One possible model for this situation would be that shown in Figure 5.11. Two parallel lines are employed, one consisting of a pure mixing-cell sequence and the second a plug flow/mixing cell sequence in series. Total feed is split between the two parallel trains. The number of mixing cells in sequence in each train would be determined by individual fit to the two peaks of the $C(t)$ or $E(t)$ response, while the residence time in the plug-flow section would be determined by the difference of the average residence times of the two peaks. Inclusion of the plug-flow section ensures a lag in the appearance of the $E(t)$ response of the top train in comparison with that of the bottom, thus producing the characteristic bimodal response.

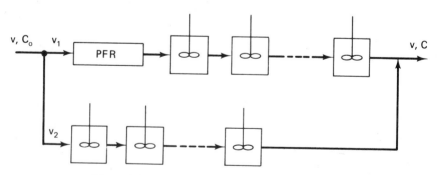

Figure 5.11 Possible combined model for channeling

[1]Actually, one could generalize this statement to . . . , a certain amount of intuition is a useful commodity. "How long halt ye between two opinions?" (1 Kings, XVIII, 21).

More complex versions of these combination models have been proposed by various workers, in particular Wolf and Resnick [D. Wolf and W. Resnick, *Ind. Eng. Chem. Fundls.*, **2**, 287 (1962)] and by Van de Vusse [J. G. Van de Vusse, *Chem. Eng. Sci.*, **17**, 507 (1962)]. These have been shown to be useful in a number of instances, but are beyond the scope of our present interests.

It should also be pointed out here that what are essentially combined models can also be used to represent micromixing effects in reaction systems. We will not attempt to present them as a separate class of mixing models here, for in essence they duplicate much of the formalism given above, although their interpretation is rather different. This is more clearly illustrated when applications to chemically reacting systems are involved, as will be discussed later.

5.4   Modeling of Nonideal Reactors

The reader very well at this point may ask what happened to the chemical reaction, since we have mostly discussed mixing models in this chapter without reference to reaction. In review of the various approaches to modeling nonideal flow effects on reactor performance, however, we find that a number of these have already been treated—although with perhaps different applications in mind. The classes of reactor models are:

1. Those making direct use of exit-age distribution information—segregated-flow models.
2. Those employing mixing-cell sequences.
3. Those employing a diffusion/dispersion term in the continuity equation.
4. Those employing combinations of ideal reactor models together with dead volume/short circuiting components.

Direct application of the residence time/exit age data has been treated in application to chemical reactor design in section 5.2a. In terms of net effect of the exit-age distribution on conversion, we wrote

$$\bar{x} = \int_0^\infty x(t)E(t)\,dt \qquad (5\text{-}1)$$

There is little further to be said here concerning the application of this method. One interesting technique, which seems not to have been exploited in the literature, is incorporation of the mixing-cell model representation for $E(t)$ in equation (5-1), as shown in section 5.2b.

Mixing-cell models of reactors were discussed extensively in chapter 4 under the guise there of the analysis of series CSTR sequences. In the following section we shall revisit some of this analysis from the point of view of modeling nonideal reactors.

a. Mixing-cell sequences as reactor models

As discussed in section 5.2b, the number of cells in sequence becomes the parameter to model deviations from the ideal PFR $F(t)$ or $E(t)$ responses. In terms of conversion alone, such deviations can be evaluated from comparison of the results computed from PFR and CSTR conversion/residence time relationships, as detailed in Tables 4.1 and 4.3. Because of the difficulties associated with deriving general (C_n/C_0) relationships for CSTR sequences involving nonlinear kinetics [cf. equation (4-80)], a generalized mixing-cell model for nonideal reactors is not conveniently obtained. The following analysis for the specific (and familiar) example of irreversible first-order kinetics should serve to illustrate the direction and magnitude of deviations from ideal behavior under steady-state conditions.[1]

Consider the PFR conversion, given by

$$x = 1 - e^{-kt_R} \tag{5-55}$$

to be the limit obtained for the ideal reactor. For the CSTR series model, conversion is given by

$$x = 1 - \frac{1}{\alpha^n} \tag{5-56}$$

where $\alpha = 1 + k\bar{t}$ and $\bar{t}$ is the (uniform) residence time per individual cell. In both equations (5-55) and (5-56) k is the first-order rate constant. Deviations from the ideal reactor conversion limit then are given by equation (5-56), in which n is determined by fit to experimental $F(t)$ or $E(t)$ results obtained for the actual reactor. The value of $\bar{t}$ is determined from the nominal PFR residence of the actual reactor by

$$\bar{t} = \frac{t_R}{n} \tag{5-57}$$

This procedure is thus far entirely analogous to that discussed for the comparisons between CSTR and PFR in chapter 4. In Figure 5.12a is given an example of the effect of a nonideal exit-age distribution on conversion in a tubular reactor modeled by equation (5-56). The PFR residence time is fixed, so in the representation of Figure 5.12a increasing deviations from ideal behavior are illustrated as n decreases (Figure 5.3). The values of δ shown on the figure are calculated from

$$\delta = \frac{(x_{PFR} - x_{model})(100)}{x_{PFR}} \tag{5-58}$$

It is seen that these deviations are really not very large unless n becomes very small. From Figure 5.3 it is clear that for $n = 20$ there are significant dif-

---

[1]To focus attention on the question of reactor modeling, we shall keep the reaction kinetics as simple as possible. Further justification for the use of first-order rate equations is provided by the fact that reactions of different order can be approximately modeled as first-order systems up to conversions on the level of 70%. "The buyer needs a hundred eyes, the seller not one" (George Herbert).

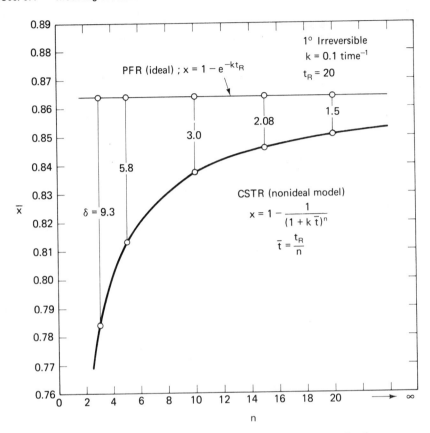

Figure 5.12a Effect of nonideal exit age distribution on conversion in a tubular flow reactor

ferences already between the residence-time and exit-age distributions of the nonideal reactor and the plug-flow case, yet the deficit in conversion is only 1.5%. The reason for this is a compensation between those elements having shorter residence times and lower conversion and those with longer residence times and higher conversion. Inspection of the response curves in Figure 5.3, however, shows that both are skewed to the left of $(t/\bar{t}) = 1$; that is, there is a weighting of elements of shorter residence time, and thus the net conversion will be reduced. Now, it may seem that we have expended a great deal of effort so far in dealing with nonideal reactor modeling to account for a tax of only 2 or 3% on net conversion. One needs only to reflect on the magnitude of most chemical, petrochemical, or petroleum process streams, however, to realize that while the percentage amounts may be relatively small, the absolute magnitudes of the quantities of materials involved can be large.

Perhaps more important than conversion in the modeling of nonideal reactors are the associated problems of selectivity and yield. Here, choose first a Type III (A $\rightarrow$ B $\rightarrow$ C) system to illustrate this. For selectivity in a PFR, Type III reaction, we had previously derived:

$$S_B(\text{III})_{\text{PFR}} = \left(\frac{k_1}{k_2 - k_1}\right) \frac{e^{-k_1 t_R} - e^{-k_2 t_R}}{1 - e^{-k_1 t_R}} \tag{5-59}$$

and for the CSTR sequence consisting of n equally sized units at the same temperature:

$$S_B(\text{III})_{\text{CSTR}} = \left(\frac{k_1}{k_1 - k_2}\right) \frac{\left(\frac{\alpha}{\beta}\right)^n - 1}{\alpha^n - 1} \tag{5-60}$$

where k_1 and k_2 are the rate constants for the first and second reaction steps, respectively, and $\alpha = 1 + k_1 \bar{t}$, $\beta = 1 + k_2 \bar{t}$. Again, in comparison between the two selectivities, $\bar{t}$ is determined from t_R according to equation (5-57). Typical results for selectivity variation in nonideal reactors according to the CSTR model are shown in Figure 5.12b for both $(k_1/k_2) > 1$ and < 1. As

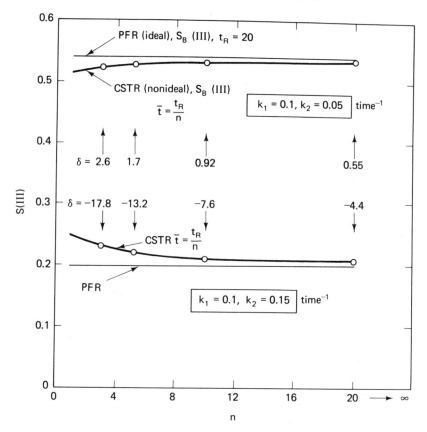

Figure 5.12b Effect of nonideal exit-age distribution on Type III selectivity in a tubular flow reactor (mixing-cell model)

in the case of comparisons of conversion, the parameter n represents the magnitude of deviation from ideality. Once more we see a decrease in selectivity with respect to the PFR limit for $(k_1/k_2) > 1$, but of relatively small magnitude. For $(k_1/k_2) < 1$ the selectivity is actually enhanced by a small amount for reasons that will be described presently.

Now let us recall the definition of selectivity:

$$\text{selectivity} = \frac{\text{moles } j \text{ produced}}{\text{moles } i \text{ reacted}}$$

As discussed previously, this defines a measure of the efficiency of a particular reaction in production of a specified intermediate or product, j, from the reactant i. Just a moment ago we were discussing, relative to conversion, that small percentage deviations might, in fact, represent large absolute amounts

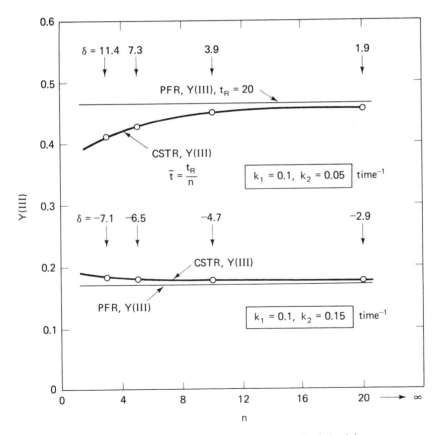

Figure 5.12c Yield results corresponding to conversion/selectivity computations

of product. In this regard the yield obtained in a complex reaction is of ultimate importance. Recall also the definition for yield:

$$\text{yield} = (\text{selectivity})(\text{conversion})$$

This is not a particularly nice relationship when we are discussing nonideal reactor modeling, since it indicates that the individual effects on conversion and selectivity are multiplied together in determination of the net loss of yield. For $S_i(\text{III}) > 1$, unfortunately, then, the decrease in yield in a nonideal reactor is greater than individual losses in conversion or selectivity; a given loss in conversion may show up as two or three times that amount in terms of the yield of a desired product. On the other hand, trends in conversion and selectivity tend to compensate each other for $S_i(\text{III}) < 1$; this is shown in Figure 5.12c for the reaction parameters treated in Figures 5.12a and b. Plotted are the product of the conversion and selectivity for the two reaction systems as a function of n. It is clear that for $S_i(\text{III}) > 1$, the δ values for yield are larger in each case than for conversion or selectivity. For example, at $n = 10$, δ (conversion) is 3.0%, δ (selectivity) is 1.6%, and δ (yield) is 4.4% ($k_1/k_2 = 2$).

Yield and selectivity comparisons in a Type III system as a function of the kinetic parameters are shown in Figure 5.12d, where the nonideal reactor model is a CSTR sequence with $n = 10$. The magnitude of the deviation from PFR performance is sensitive to the relative values of k_1 and k_2, increasing as the intrinsic selectivity $[S_i(\text{III}) = k_1/k_2]$ increases. There exists a region for $S_i(\text{III}) < 1$ where the selectivity for the nonideal reactor is actually slightly greater than the PFR value, while the yields are essentially the same. This is the result of compensation between elements of shorter and longer residence times. Here $k_1 < k_2$, so that longer residence times tend to promote the faster reaction B $\xrightarrow{k_2}$ C, removing the desired intermediate. Conversely, shorter residence times do not provide as much opportunity for the intermediate to react away and, as we see here, for this range of k_1 and k_2 the skew toward shorter residence time in the nonideal reactor is sufficient to promote enhancement of the selectivity.

The analysis here should convey the message that generalizations concerning selectivity or yield performance in nonideal reactors with reference to an ideal model are dangerous; conversion, on the other hand, is perhaps somewhat more predictable. We may normally expect modest taxes on conversion as the result of nonideal exit-age distributions; if the reaction system involves selectivity/yield functions these will also be influenced by the exit-age distribution. In many instances we can expect this to be reflected as a decrease in yield and selectivity, but there are possible interactions between the reactor model exit-age distribution and the kinetic parameters that can force this deviation in the opposite direction. Keep in mind that the comparisons being offered here are not analogous to those for PFR–CSTR Type III selectivities

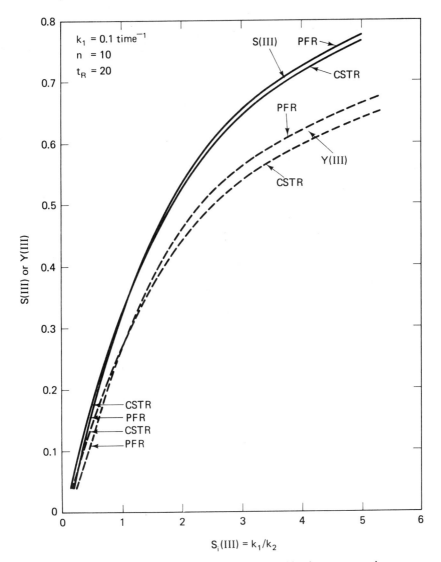

Figure 5.12d Selectivity and yield variations with kinetic parameters in a nonideal reactor

given in chapter 4, which were based on the premise of equal conversion in the two reactor types.

To complete this discussion of conversion and yields in nonideal reactors, let us consider a similar set of illustrations for a Type II system. Here

$$A \begin{array}{c} \overset{k_1}{\nearrow} B \\ \underset{k_2}{\searrow} C \end{array}$$

and the pertinent relationships are, for the PFR model,

$$x = 1 - e^{-(k_1 + k_2)t_R} \qquad (5\text{-}61)$$

$$S_B(II)_{PFR} = \frac{k_1}{k_1 + k_2} \qquad (5\text{-}62)$$

and for the CSTR sequence,

$$x = 1 - \gamma^{-n} \qquad (5\text{-}63)$$

$$S_B(II)_{CSTR} = \frac{k_1}{k_1 + k_2} \qquad (5\text{-}64)$$

where $\gamma = 1 + (k_1 + k_2)\bar{t}$. In both cases yield is calculated as before from the product of conversion and selectivity. Now it is apparent here that since the selectivity in both cases is the same and is constant, the effects of nonideality on yield in the Type II system will be in the same direction as on conversion, and in fact will be of the same magnitude. This is illustrated in Figures 5.13a and b for an example using the same kinetic parameters just employed for the Type III case.

A comparison of conversion and yield for Type II in terms of the kinetic parameters using a nonideal reactor model ($n = 10$) is shown in Figure 5.13c.

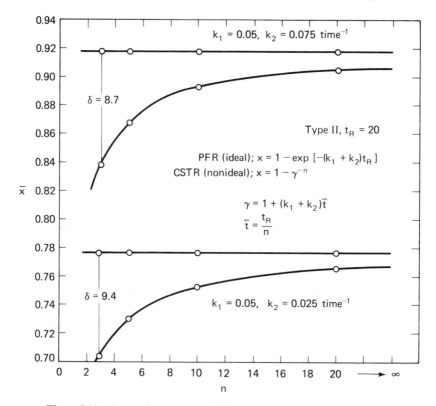

Figure 5.13a Conversion in a nonideal reactor for a Type II reaction system

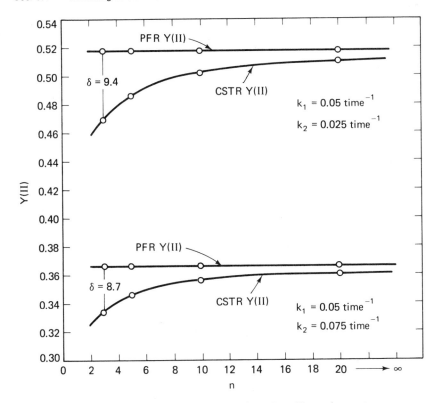

Figure 5.13b Yield in a nonideal reactor for a Type II reaction system

Here as the value of (k_1/k_2) decreases, yield and conversion in the nonideal reactor approach the ideal value; in this case this is a limiting value, owing to the equality of the selectivities in the two reactor models.

Use of the CSTR sequence as a model for nonideal reactors has been criticized on the basis of the model's lack of certain aspects of physical reality, such as the absence of backward communication between the individual mixing-cell units. Nonetheless, the mathematical simplicity of the approach makes it extremely attractive, particularly for systems with complex kinetics, nonisothermal effects, or other complicating factors. Indeed, in this respect computer-time requirements may dictate the selection of a particular modeling approach, a feature in which the mixing-cells model possesses a distinct advantage over alternative approaches.

b. Axial dispersion reactor models

A nonideal reactor model based on the axial dispersion (one-dimensional) equation, (5-20), is readily written down following the procedure used in deriving the PFR relationship, equation (4-40). Hence, equation (5-20)

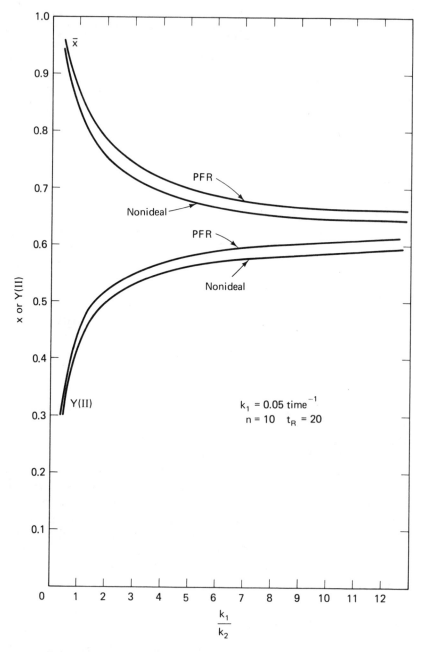

Figure 5.13c Conversion and yield, Type II, in a nonideal reactor

becomes, for steady-state conditions,

$$D\frac{d^2C}{dz^2} - u\frac{dC}{dz} - (-r) = 0 \tag{5-65}$$

where C now refers to the concentration of a reactant or product of interest. Again following the convention adopted for the PFR equation, $-r$ is a net positive quantity if C refers to reactant, negative if C refers to product.

We have previously pointed out that use of the dispersion model changes the reactor analysis from an initial-value to a boundary-value problem; thus, it is necessary to worry some about the form of boundary conditions to use for equation (5-65). This is illustrated in Figure 5.14, where several different

1.   Open–Open

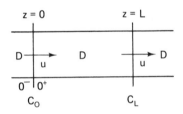

At $z = 0$: $uC_0 - D\left(\dfrac{dC}{dz}\right)_{0^-} = -D\left(\dfrac{dC}{dz}\right)_{0^+}$

$C_{0^-} = C_{0^+}$

At $z = L$: $C = C_L$

3.   Open–Closed

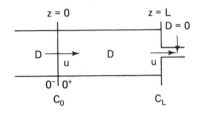

At $z = 0$: $uC_0 - D\left(\dfrac{dC}{dz}\right)_{0^-} = -D\left(\dfrac{dC}{dz}\right)_{0^+}$

$C_{0^-} = C_{0^+}$

At $z = L$: $\left(\dfrac{dC}{dz}\right)_L = 0$

2.   Closed–Closed

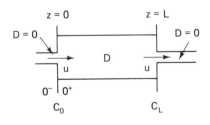

At $z = 0$: $uC_{0^+} - uC_0 = D\left(\dfrac{dC}{dz}\right)_{0^+}$

At $z = L$: $\left(\dfrac{dC}{dz}\right)_L = 0$

4.   Closed–Open

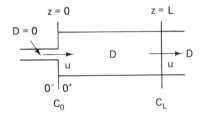

At $z = 0$: $uC_{0^-} - uC_0 = D\left(\dfrac{dC}{dz}\right)_{0^+}$

At $z = L$: $C = C_L$

Figure 5.14 Different inlet/outlet configurations with boundary conditions for one-dimensional dispersion reactor model

configurations of inlet and outlet conditions are shown. In the *open–open configuration*, for example, we consider that dispersion in both fore and aft sections, outside the reactor proper, is possible. In the *closed–closed configuration* only convective transport occurs external to the reactor. Intermediate cases such as open–closed and closed–open configurations might also more closely conform to particular instances of flow patterns associated with various types of reactor design. Obviously, the complexity of the solution to equation (5-65) is going to depend on the complexity of the boundary conditions employed, so often some compromise may be required between the realities of nonideal flow patterns and realistic boundary condition approximations. Three sets of conditions have been employed in most of the literature dealing with one-dimensional dispersion reactor models. These are:

TYPE A: $(z = 0)$ $uC_0 = uC_{0^+} - D\left(\dfrac{dC}{dz}\right)_{0^+}$

$(z = L)$ $\left(\dfrac{dC}{dz}\right) = 0$ (5-66)

TYPE B: $(z = 0)$ $uC_0 = uC_{0^+}$

$(z = L)$ $\left(\dfrac{dC}{dz}\right) = 0$ (5-67)

TYPE C: $(z = 0)$ $uC_0 = uC_{0^+}$

$(z \rightarrow \infty)$ $\lim C(z) = 0$ (5-68)

Typical references for the uses of these three types are provided in the writings of Danckwerts [P. V. Danckwerts, *Chem. Eng. Sci.*, **2**, 1 (1953)], Hulburt [H. M. Hulburt, *Ind. Eng. Chem.*, **36**, 1012 (1944)], and Levenspiel and Smith [O. Levenspiel and W. K. Smith, *Chem. Eng. Sci.*, **6**, 227 (1957)] for A, B and C, respectively. The notation employed in equations (5-66) to (5-68) is illustrated in Figure 5.14. It is seen that the Type A conditions listed above correspond to the closed–closed configuration, case 2 in Figure 5.14. The reasoning involved in writing such a set of boundary conditions for this case is roughly as follows. At the entrance to the reactor the concentration is established by convective transport from the region $z < 0$ across the plane $z = 0$, which must be balanced by the combination of convective transport and dispersion occurring within the reactor ($z = 0^+$). No dispersion occurs for $z < 0$ so this term is absent from the left side of equation (5-66) for $z = 0$. A similar balance may be written for the exit of the reactor, with convection plus dispersion at $z = L$ balanced by convection at $z = L^+$, so that

$$uC_{\text{out}} = uC_L - D\left(\frac{dC}{dz}\right)_L$$

Now if $(dC/dz)_L$ is negative, the concentration in the exit stream must be greater than that within the reactor, whereas if $(dC/dz)_L$ is positive, the concentration of reactant would have to pass through a minimum somewhere in the reactor. Neither one of these alternatives is physically reasonable, so we are left with

$$uC_{\text{out}} = uC_L$$

or

$$\left(\frac{dC}{dz}\right)_L = 0$$

To evaluate which among the alternative conditions presented by Types A, B, and C is most correct in an absolute sense, we might establish as a criterion that the solution with the correct boundary conditions should reduce to the PFR limit for $D \longrightarrow 0$, and to the CSTR limit for $D \longrightarrow \infty$.[1] In an extensive analysis, Fan and Ahn [L-T. Fan and Y-K. Ahn, *Ind. Eng. Chem. Proc. Design Devel.* **1**, 190 (1962)] have shown that A, as proposed by Danckwerts, is the only one of the three that yields this proper limiting behavior. Let us illustrate this for first-order irreversible kinetics, for which an analytical solution is possible.[2] It is first convenient to render equations (5-65) and (5-66) into nondimensional form by substitution of the following dimensionless variables:

$$f = \frac{C}{C_0} = 1 - x \qquad \zeta = \frac{z}{L}$$

$$N_{\text{Pe}} = \frac{Lu}{D} \qquad R' = \frac{kL}{u} = \frac{L}{uC}(-r)$$

whence

$$\frac{1}{N_{\text{Pe}}} \frac{d^2 f}{d\zeta^2} - \frac{df}{d\zeta} - R'f = 0 \tag{5-69}$$

$$(\zeta = 0): \qquad \frac{df(0^+)}{d\zeta} = N_{\text{Pe}}[f(0^+) - 1] \tag{5-70}$$

$$(\zeta = 1): \qquad \frac{df(1)}{d\zeta} = 0$$

The solution obtained by Danckwerts for f as a function of ζ is

$$f(\zeta) = \exp\left(\frac{N_{\text{Pe}}}{2}\zeta\right)$$

$$\cdot \left[\frac{2(1 + \beta) \exp\left[\frac{N_{\text{Pe}}\beta}{2}(1 - \zeta)\right] - 2(1 - \beta) \exp\left[\frac{N_{\text{Pe}}\beta}{2}(\zeta - 1)\right]}{(1 + \beta)^2 \exp\left(\frac{N_{\text{Pe}}}{2}\beta\right) - (1 - \beta)^2 \exp\left(\frac{-N_{\text{Pe}}}{2}\beta\right)}\right] \tag{5-71}$$

---

[1] The establishment of these bounding limits on D for ideal flow cases can be conveniently visualized physically from equation (5-65). For example, as $D \longrightarrow \infty$ for the first term to be bounded, then $(d^2C/dz^2) \longrightarrow 0$, which requires that the concentration be uniform with z. "A mighty maze! but not without a plan" (Alexander Pope).

[2] Results for other kinetics are reported by Fan and Balie [L-T. Fan and R. C. Balie, *Chem. Eng. Sci.*, **13**, 63 (1960)] and by Burghardt and Zaleski [A. Burghardt and T. Zaleski, *Chem. Eng. Sci.*, **23**, 575 (1968)].

where

$$\beta = \left(1 + \frac{4R'}{N_{\mathrm{Pe}}}\right)^{1/2}$$

Conversion at the reactor exit is obtained from evaluation of $1 - f$ from equation (5-71) for $\zeta = 1$:

$$x(1) = 1 - \frac{4\beta}{(1+\beta)^2 \exp\left[-\frac{N_{\mathrm{Pe}}}{2}(1-\beta)\right] - (1-\beta)^2 \exp\left[-\frac{N_{\mathrm{Pe}}}{2}(1+\beta)\right]} \tag{5-72}$$

Now let us examine the limiting forms of equation (5-72) for $D \to 0$ and $D \to \infty$. As $D \to 0$, $4R'/N_{\mathrm{Pe}}$ must also approach zero, so that $\beta \to 1$ and

$$\exp\left[-\frac{N_{\mathrm{Pe}}}{2}(1-\beta)\right] \longrightarrow \exp[-R']$$

from equation (5-72) if we make the approximation that

$$1 - \left(1 + \frac{4kD}{u^2}\right)^{1/2} \approx \frac{2kD}{u^2}$$

for small D. Then

$$x(1) = 1 - \exp[-R'] \tag{5-73}$$

which is the proper PFR result for these kinetics. For $D \to \infty$ we can expand the exponential terms in the denominator linearly:

$$x(1) = 1 - \frac{4\beta}{(1+\beta)^2\left(1 - \frac{N_{\mathrm{Pe}}}{2} + \frac{N_{\mathrm{Pe}}\beta}{2}\right) - (1-\beta)^2\left(1 + \frac{N_{\mathrm{Pe}}}{2} - \frac{N_{\mathrm{Pe}}\beta}{2}\right)}$$

Algebraic simplification of the expression yields

$$x(1) = \frac{1}{1 + 1/R'} \tag{5-74}$$

which is the proper CSTR limit.

For very small but not vanishing values of D, the following result can be obtained:

$$x(1) = 1 - \left(1 + \frac{k^2 DL}{u^3}\right)\exp(-R') \tag{5-75}$$

so the deviation from PFR behavior is given by the magnitude of $(1 + k^2 DL/u^3)$. Hence, a criterion for negligible dispersion effects is given by

$$\frac{k^2 DL}{u^3} \ll 1 \tag{5-76a}$$

or

$$\frac{(\ell n\, f)^2}{N_{\mathrm{Pe}}} \ll 1 \tag{5-76b}$$

The solutions of equation (5-69) for the other types of boundary conditions, B and C, are given in Table 5.2. The result for Type B is almost as

TABLE 5.2

SOLUTIONS TO THE ONE-DIMENSIONAL DISPERSION MODEL FOR TYPES B AND C BOUNDARY CONDITIONS, IRREVERSIBLE FIRST-ORDER KINETICS

Type B

$$f(\zeta) = \frac{(1+\beta)\exp\left[\frac{N_{Pe}}{2}(1+\beta) + \frac{N_{Pe}}{2}(1-\beta)\zeta\right] - (1-\beta)\exp\left[\frac{N_{Pe}}{2}(1+\beta)\zeta + \frac{N_{Pe}}{2}(1-\beta)\right]}{(1+\beta)\exp\left[\frac{N_{Pe}}{2}(1+\beta)\right] - (1-\beta)\exp\left[\frac{N_{Pe}}{2}(1-\beta)\right]}$$

Type C

$$f(\zeta) = \exp\left[\frac{N_{Pe}}{2}(1-\beta)\zeta\right]$$

$$\beta = \left(1 + \frac{4R'}{N_{Pe}}\right)^{1/2}$$

complicated as for Type A; however, the Type C result is a simple exponential form which is obviously easier to use for computation than equation (5-71) or (5-72). It is of interest, then, to determine when significant deviations may be expected from the Danckwerts solution, and such a comparison has been reported by Fan and Ahn. In Figure 5.15 are plotted deviations of the solutions obtained with Types B and C boundary conditions from Type A at entrance, midsection, and exit of the reactor. Such deviations can be expressed entirely in terms of the parameters R' and N_{Pe}, as shown in the figure. The differences are most pronounced for small values of R' and N_{Pe}, a condition which we could visualize qualitatively as for a sluggish reaction in a reactor with large nonideal flow effects. Also evident from the figure is that Type B conditions result in larger deviations than those for Type C.

A quick and convenient method for computing conversions using the axial dispersion Type A model is to estimate the value using the simple exponential of Type C and then apply the correction indicated for the reactor exit on Figure 5.15 for the particular values of N_{Pe} and R' involved. Details on numerical methods for the solution of these equations are provided in chapter 6.

The magnitude of nonideality in the dispersion model is given by the value of N_{Pe}, as we have discussed previously. In Figure 5.16 are presented the results of calculations for two of the axial dispersion models in terms of N_{Pe} as compared with the PFR result. The value of R' employed in this calculation corresponds to the kinetic parameters employed for the CSTR model comparison given in Figure 5.12a, so the results in the two figures are directly comparable. As an example, consider the conversion of 0.833 deter-

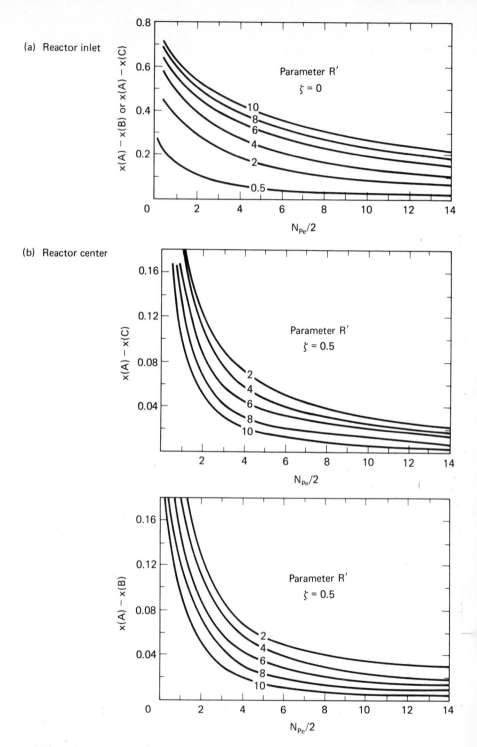

Figure 5.15 Differences in conversion computed from boundary conditions A, B, and C for reactor inlet, center, and exit (reprinted from L. T. Fan and Y. K. Ahn, *Ind. Eng. Chem. Process Design Devel.*, **1**, 190 (1962); copyright by the American Chemical Society)

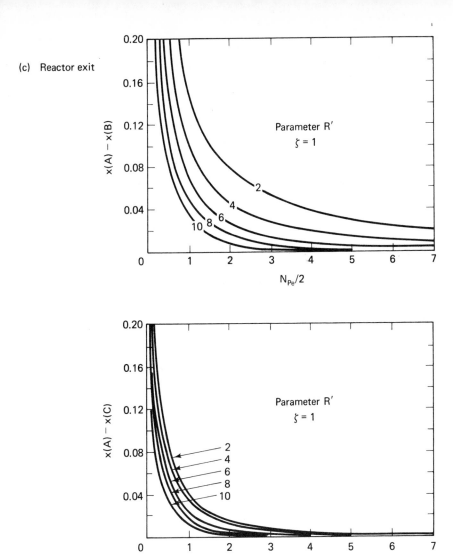

(c) Reactor exit

Parameter R'

$\zeta = 1$

Figure 5.15 (cont.)

mined from the axial dispersion model for $N_{Pe} = 14$. The same conversion level corresponds to $n \approx 9$ from Figure 5.12a, indicating that the residence-time distributions generated by the two models should be equivalent. This comparison of $F(t)$ for the CSTR series model, $n = 9$, and the axial dispersion model, $N_{Pe} = 14$, is shown in Figure 5.17.

The most general solution to the one dimensional axial dispersion model has been given by Wehner and Wilhelm [J. F. Wehner and R. H. Wilhelm, *Chem. Eng. Sci.*, **6**, 89 (1956)]. It is formulated in such a way that a variety of the cases illustrated in Figure 5.14 can be obtained as particular solutions to the general model. In terms of the boundary conditions, we make no assumptions concerning the role of dispersion in the sections fore and aft

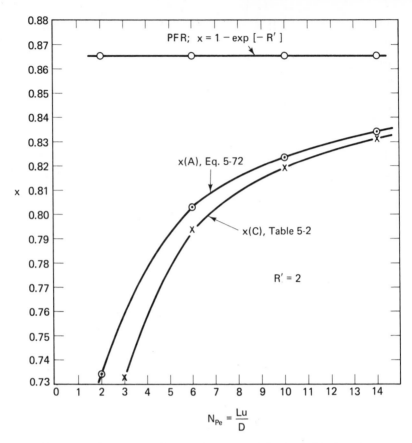

Figure 5.16 Some conversion results computed by the axial dispersion model

of the reactor proper; thus, the overall system is visualized as shown in Figure 5.18. The end sections extend from $-\infty$ to $+\infty$, with the reaction system located from $0 \leq \zeta \leq 1$. The overall boundary conditions then apply at $\pm\infty$, and a total mass flux balance including both dispersion and convection can be written between sections a and b and b and c. Using the nondimensional variables defined for equation (5-69), the dispersion equations for the three sections are

$$\text{a)} \quad \frac{1}{(N_{\text{Pe}})_a} \frac{d^2f}{d\zeta^2} - \frac{df}{d\zeta} = 0; \zeta \leq 0 \tag{5-77}$$

$$\text{b)} \quad \frac{1}{(N_{\text{Pe}})_b} \frac{d^2f}{d\zeta^2} - \frac{df}{d\zeta} - R'f = 0; 0 \leq \zeta \leq 1 \tag{5-78}$$

$$\text{c)} \quad \frac{1}{(N_{\text{Pe}})_c} \frac{d^2f}{d\zeta^2} - \frac{df}{d\zeta} = 0; \zeta \geq 1 \tag{5-79}$$

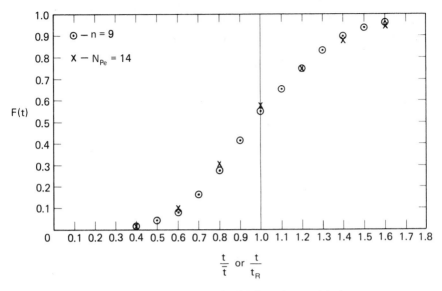

Figure 5.17 $F(t)$ curves for CSTR and axial dispersion models demonstrating identical effects on conversion in a first-order reaction

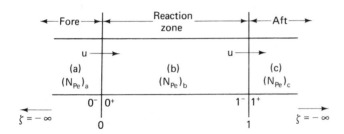

Figure 5.18 The axial dispersion model for the analysis of Wehner and Wilhelm

There are now three boundary value problems involved in description of the overall reactor, so six boundary conditions are required. These are

(1) $f(-\infty) = 1$ $\zeta = -\infty$

(2) $f(0^-) - \dfrac{1}{(N_{Pe})_a} \dfrac{df(0^-)}{d\zeta} = f(0^+) - \dfrac{1}{(N_{Pe})_b} \dfrac{df(0^+)}{d\zeta}$ $\zeta = 0$

(3) $f(0^-) = f(0^+)$ $\zeta = 0$

(4) $f(1^-) - \dfrac{1}{(N_{Pe})_b} \dfrac{df(1^-)}{d\zeta} = f(1^+) - \dfrac{1}{(N_{Pe})_c} \dfrac{df(1^+)}{d\zeta}$ $\zeta = 1$

(5) $f(1^-) = f(1^+)$ $\zeta = 1$

(6) $f(\infty) = \text{finite}$ $\zeta = \infty$

$$(5\text{-}80)$$

The general solutions obtained for the three reactor sections with these boundary conditions are

(a) $\dfrac{1-f}{1-f(0)} = \exp\left[(N_{\text{Pe}})_a \zeta\right]$ $\qquad\qquad\qquad \zeta \leq 0$ $\qquad$ (5-81)

(b) $\qquad f = g_0 \exp\left[\dfrac{(N_{\text{Pe}})_b \zeta}{2}\right]\left\{(1+\beta)\exp\left[\dfrac{\beta(N_{\text{Pe}})_b(1-\zeta)}{2}\right]\right.$

$\qquad\qquad\qquad \left. - (1-\beta)\exp\left[\dfrac{\beta(N_{\text{Pe}})_b(\zeta-1)}{2}\right]\right\}$ $\qquad 0 \leq \zeta \leq 1$ $\qquad$ (5-82)

(c) $\qquad f = f(1) = 2\beta g_0 \exp\left[\dfrac{(N_{\text{Pe}})_b}{2}\right]$ $\qquad\qquad \zeta \geq 1$ $\qquad$ (5-83)

where

$$\beta = \left[1 + \dfrac{4R'}{(N_{\text{Pe}})_b}\right]^{1/2}$$

$$g_0 = 2\left\{(1+\beta)^2 \exp\left[\beta\dfrac{(N_{\text{Pe}})_b}{2}\right] - (1-\beta)^2 \exp\left[-\beta\dfrac{(N_{\text{Pe}})_b}{2}\right]\right\}^{-1} \quad (5\text{-}84)$$

$$f(0) = g_0\left\{(1+\beta)\exp\left[\beta\dfrac{(N_{\text{Pe}})_b}{2}\right] - (1-\beta)\exp\left[-\beta\dfrac{(N_{\text{Pe}})_b}{2}\right]\right\} \quad (5\text{-}85)$$

It is interesting to note that the $f - \zeta$ profile in the fore section depends on the a–b boundary concentration, $f(0)$, which in turn is a function only of the parameters in the reaction section, $(N_{\text{Pe}})_b$ and R', while the profile in the reaction section does not depend on the parameters of the fore section. The magnitude of $(N_{\text{Pe}})_a$ controls the shape of the concentration profile in the vicinity of the a–b boundary. As $(N_{\text{Pe}})_a$ increases, the profile in the vicinity of $z = 0$ becomes increasingly steep and approaches in the limit the Type A boundary condition of Danckwerts.

Reference was made earlier to some specific numerical studies which have been reported for the axial dispersion model employing rate expressions other than first-order. Some results given by Fan and Balie for half-, second-, and third-order kinetics are illustrated in Figure 5.19. In these results the parameter R'_n is defined as

$$R'_n = \dfrac{2knL}{u}C_0^{n-1} \qquad (5\text{-}86)$$

where n is the order of reaction. These conversions can be compared with the corresponding PFR values, which are conveniently computed from

$$x = 1 - \left[1 - \dfrac{R'_n(1-n)}{2n}\zeta\right]^{1/(1-n)} \qquad (5\text{-}87)$$

It is apparent from the shapes of the curves given in Figure 5.19 that interpolation or extrapolation for various orders and N_{Pe} is at best difficult and very risky. Thus, separate numerical solutions will in general be required for

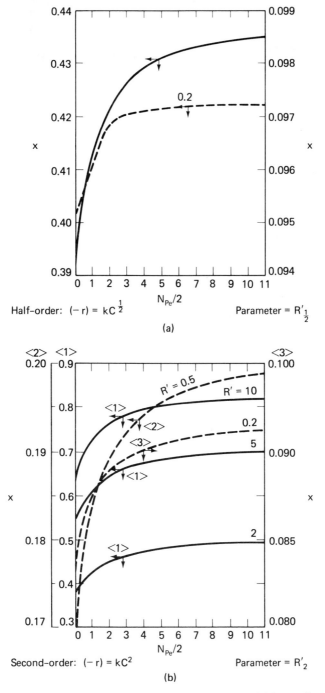

0.44 / 0.099
0.43 / 0.098
0.43 ← ↓
0.2 ↓
0.42 / 0.097
0.41 / 0.096
0.40 / 0.095
0.39 / 0.094

$N_{Pe}/2$

Half-order: $(-r) = kC^{\frac{1}{2}}$ Parameter = $R'_{\frac{1}{2}}$

(a)

$R' = 0.5$ $R' = 10$
0.2
5
2

Second-order: $(-r) = kC^2$ Parameter = R'_2

(b)

Figure 5.19 Conversion according to axial dispersion model for nonlinear kinetic expressions (after L. T. Fan and R. C. Balie, *Chem. Eng. Sci.*, **13**, 63 (1960); with permission of Pergamon Press, Ltd., London)

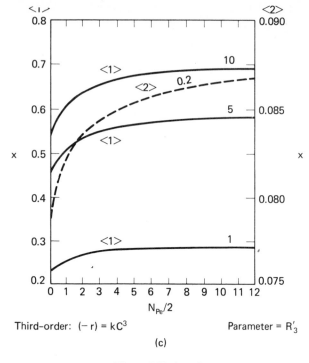

Third-order: $(-r) = kC^3$ Parameter $= R'_3$

(c)

Figure 5.19 (cont.)

nonlinear kinetics. In this regard, modeling with the CSTR sequence is much easier and more flexible. A similar comment applies to selectivity problems, particularly for Type III systems.

c. Combined models

Conversion equations for the combined models are easily derived by the same general procedures demonstrated in the mass-balance equations for CSTR and PFR systems. Consider the CSTR with short-circuiting illustrated in Table 5.1. If we write, for steady-state conditions, a mass balance around the vessel of volume $\bar{V}$, for first-order kinetics:

$$v_1 C_0 = v_1 C_1 + k C_1 \bar{V}$$

or

$$\frac{C_1}{C_0} = \frac{1}{1 + k\bar{t}_1} \tag{5-88}$$

where $\bar{t}_1 = \bar{V}/v_1$. Now, at the exit the short-circuit stream combines with the reactor effluent so that

$$vC = v_1 C_1 + v_2 C_0 \tag{5-89}$$

Substituting for C_1 from equation (5-88) and rearranging yields

$$\frac{C}{C_0} = \frac{v_1/v}{1 + k\bar{t}_1} + \frac{v_2}{v}$$

and since $\bar{t} = (v_1/v)\bar{t}_1$ according to the definitions employed in Table 5.1, we obtain

$$\frac{C}{C_0} = \frac{v_1/v}{1 + (v/v_1)k\bar{t}} + \frac{v_2}{v} \tag{5-90}$$

Similar relationships for the steady-state conversions according to the combined models given previously are listed in Table 5.3.

TABLE 5.3

SOLUTIONS TO EXAMPLE COMBINED MODELS
(NOTATION AS IN TABLE 5.1)

1. Plug flow with dead space:

$$\frac{C}{C_0} = \exp\left(-fkt_R\right)$$

2. CSTR with dead space:

$$\frac{C}{C_0} = \frac{1}{1 + fk\bar{t}}$$

3. CSTR with short-circuiting:

$$\frac{C}{C_0} = \frac{v_1/v}{1 + (v/v_1)k\bar{t}} + \frac{v_2}{v}$$

4. PFR–CSTR sequence:

$$\frac{C}{C_0} = \frac{\exp\left[-f_1 k(\bar{t} + t_R)\right]}{1 + f_2 k(\bar{t} + t_R)}$$

5. CSTR with short-circuiting and dead volume:

$$\frac{C}{C_0} = \frac{v_1/v}{1 + f(v/v_1)k\bar{t}} + \frac{v_2}{v}$$

d. Mixing-cell, axial dispersion, and combined models: the unsteady state

The models as presented and analyzed in the three previous sections are steady state and, as such, yield no information as to how the state was attained. Reactor startup and shutdown operations, however, are often an important, even critical, factor in actual processing applications. This is particularly so in applications of stirred tank reactors, in which one reactor or reactor series may be used for a variety of reactions on some time-scheduled basis. Operation during startup would normally involve the production of products which do not meet quality standards or other specifications, so it is naturally of interest to be able to analyze such operation.

The mixing models presented in section 5.2 provide the basis for unsteady-state reactor analysis. These mixing models are all unsteady-state ones, since they are devised for the interpretation of the responses, $F(t)$ or $E(t)$. For the CSTR sequence, the general mass balance with chemical reaction becomes

$$\frac{dC_n}{dt} + \frac{C_n}{\bar{t}} + (-r) = \frac{C_{n-1}}{\bar{t}}$$

which for an irreversible first-order reaction in a single unit becomes

$$\frac{dC_{A_1}}{dt} + C_{A_1}\left(\frac{1+k\bar{t}}{\bar{t}}\right) = \frac{C_A(t)}{\bar{t}} \tag{5-91}$$

where $C_A(t)$ is the time-dependent feed concentration. This equation may be solved conveniently via Laplace transforms for a step change from an initial value, say C_{A_0}, to a new value, $C_{A_0} + \Delta C_{A_0}$. The result is

$$C_{A_1}(t) = \frac{C_{A_0} + \Delta C_{A_0}}{1+k\bar{t}} + \exp\left[(1+k\bar{t})\frac{t}{\bar{t}}\left(A_1 - \frac{C_{A_0}+\Delta C_{A_0}}{1+k\bar{t}}\right)\right] \tag{5-92}$$

where A_1 is the initial value of C_{A_1} and would be obtained from the steady-state solution. The first-order case may be generalized for the n-unit sequence, provided that equal temperatures and volumes pertain for each unit, to the following:

$$C_{A_n} = \frac{C_{A_0}}{Q^n} + e^{-Qt/\bar{t}}\left\{A_n + A_{n-1}\left(\frac{t}{\bar{t}}\right) + \cdots + \frac{A_1}{(n-1)!}\left(\frac{t}{\bar{t}}\right)^{n-1}\right.$$
$$\left. - \frac{C_{A_0}}{Q^n}\left[1 + \frac{Qt}{\bar{t}} + \cdots + \frac{1}{(n-1)!}\left(\frac{Qt}{\bar{t}}\right)^{n-1}\right]\right\} \tag{5-93}$$

where C_{A_n} at time zero is given by A_n, and $Q = 1 + k\bar{t}$.

An approximate solution for second-order irreversible kinetics in a single CSTR can be obtained by linearization of the rate term. The solution is the same as equation (5-92), except the term $(1 + k\bar{t})$ is replaced by $(1 + 2k\bar{t}C_{A_0})$, where C_{A_0} is the initial value of feed concentration. This result depends on the fact that $C_{A_1}(t)$ does not differ greatly from $C_{A_1}(0)$, or more specifically that

$$2C_{A_1}(0)[C_{A_1}(t) - C_{A_1}(0)] \approx [C_{A_1}(t)^2 - C_{A_1}(0)^2]$$

In many cases this can prove to be a very restrictive assumption, so that in general the evaluation of transients for CSTR sequences involving non-first-order kinetics is a numerical problem. Various other cases of unsteady-state CSTR operation are summarized in the papers of Piret and coworkers cited in chapter 4.

A transient solution for step-change concentration perturbation in the axial dispersion model, using the Type A boundary conditions, has been given by Fan and Ahn [L. T. Fan and Y. K. Ahn, *Chem. Eng. Progr. Symp. Ser.*, **46**(59), 91 (1963)]. For a step change of magnitude C_{A_0},

$$\frac{C(t)}{C_{A_0}} = 1 - 4 \sum_{n=1}^{\infty} \frac{M \delta_n (M \sin \delta_n + \delta_n \cos \delta_n)}{(M^2 + 2M + \delta_n^2)(M^2 + \delta_n^2 + 2MR')}$$
$$\cdot \exp\left[M - \left(\frac{M^2 + \delta_n^2 + 2MR'}{2M} \right)\left(\frac{t}{t_R} \right) \right]$$

(5-94)

where

$$M = \frac{N_{Pe}}{2}$$

$$\delta_n = n\text{th root of } \cot \delta = \frac{1}{2}\left(\frac{\delta}{M} - \frac{M}{\delta} \right)$$

$$R' = k t_R$$

Some examples of transients computed from equation (5-94) are shown in Figure 5.20 for $R' = 0.1$. The complexity of equation (5-94) is such that even though the solution exists in analytical form, numerical evaluation is most readily done via machine computation.

A comparison of these transient forms for first-order kinetics, equation (5-93) for the CSTR sequence and (5-94) for the axial dispersion model, on

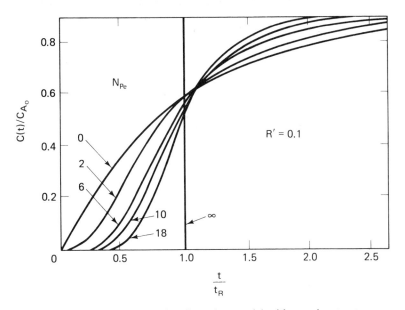

Figure 5.20 Response of axial dispersion model with reaction to step-function concentration input, $0 \longrightarrow C_{A_0}$

the basis of their steady-state equivalence in terms of n and N_{Pe} is suggested in the exercises. The basic question to be answered is whether the equivalence of the two models in the steady state is sufficient to ensure their equivalence in modeling the unsteady-state (startup) operation.

The transient forms of the combined models, corresponding to those shown in Tables 5.1 and 5.3, are given in Table 5.4.

TABLE 5.4

UNSTEADY-STATE SOLUTIONS FOR COMBINED MODELS,
FIRST-ORDER REACTION

1. Plug flow with dead space:

$$\frac{C(t)}{C_0} = \exp\left(-fkt_R\right) \delta\left(\frac{t}{t_R} - f\right)$$

2. CSTR with dead space:

$$\frac{C(t)}{C_0} = \frac{1}{1 + fk\bar{t}} - \frac{1}{1 + fk\bar{t}} \exp\left[-(1 + fk\bar{t})\frac{t}{f\bar{t}}\right]$$

3. CSTR with short-circuiting:

$$\frac{C(t)}{C_0} = \frac{(v_1/v)^2}{(v_1/v) + k\bar{t}}\left\{1 - \exp\left[-\left(k\bar{t} + \frac{v_1}{v}\right)\frac{t}{\bar{t}}\right] + \frac{v_2}{v}\delta\left(\frac{t}{\bar{t}} = 0\right)\right\}$$

4. PFR–CSTR sequence:

$$\frac{C(t)}{C_0} = \left\{\frac{\exp\left[-f_1k(\bar{t} + t_R)\right]}{1 + f_2k(\bar{t} + t_R)} - \frac{\exp\left\{f_1/f_2 - [1 + f_2k(\bar{t} + t_R)]\right\}t/f_2\bar{t}}{1 + f_2k(\bar{t} + t_R)}\right\} \delta\left(\frac{t}{\bar{t}} - f_1\right)$$

5. CSTR with short-circuiting and dead volume:

$$\frac{C(t)}{C_0} = \frac{(v_1/v) + (fv_1^2/v)k\bar{t}}{(v_1/v) + fk\bar{t}} - \frac{(v_1/v)^2}{(v_1/v) + fk\bar{t}} \exp\left[-\left(\frac{v_1}{v} + fk\bar{t}\right)\frac{t}{f\bar{t}}\right]$$

EXERCISES FOR CHAPTER 5

Section 5.2a

1. Compare the result of integration of the segregated-flow model for first-order reaction in a laminar flow reactor with the result presented in chapter 4.
2. Compare the conversion obtained for the second-order reaction $2A \longrightarrow C + D$ in a single CSTR with that determined from the segregated flow result of equation (5-3). The parameter $kC_{A_0}\bar{t} = 0.079$ liter/g mole. Is this result consistent with the postulated effects of micromixing on reactions of order greater than unity?
3. Using the residence-time distribution function for a laminar flow reactor, compare the yield of B for the reaction

$$A \xrightarrow{k_1} B \xrightarrow{k_2} C$$

with that obtained in a PFR of the same average residence time. Initially, there is no B or C present in the system, $k_1 = 2k_2$, and $k_2 \bar{t} = 1$.

Section 5.2b

4. (a) What is the ratio of the $F(t)$ function for two CSTR sequences, one of five units and one of nine units, at $(t/\bar{t})$ of 0.5, 1.0, and 1.5?
   (b) Repeat for $E(t)$.
5. Derive an expression for the response of an n-unit CSTR sequence to a step-function decrease in inlet concentration.
6. Determine the proper CSTR mixing model to fit the response data of problem 3, chapter 4.

Section 5.2c

7. Determine the proper dispersion model to fit the response data of problem 3, chapter 4.
8. What is the ratio of the $F(t)$ and $E(t)$ functions for two dispersion models, of $N_{Pe} = 15$ and $N_{Pe} = 5$, at $t/t_R = 0.5$, 1.0, and 1.5?
9. (a) Determine N_{Pe} according to the Taylor–Aris model for a component of a solution in laminar flow in a tube of 0.5-cm radius at an average velocity of 10 m/hr. The molecular diffusivity is 2.3×10^{-5} cm^2/sec.
   (b) What would the $F(t)$ curve look like for this situation?

Section 5.2d

10. Demonstrate by calculation that the results obtained in problems 6 and 7 indeed produce the exit-age distribution of problem 3, chapter 4.

Section 5.3

11. The results illustrated in Figure 5.21 were obtained for $F(t)$ in a given system. Determine an appropriate combined mixing model and the associated parameters that will provide a reasonable representation of this response. [Hint: The mixing may be more easily visualized in terms of $E(t)$ determined from these data.]

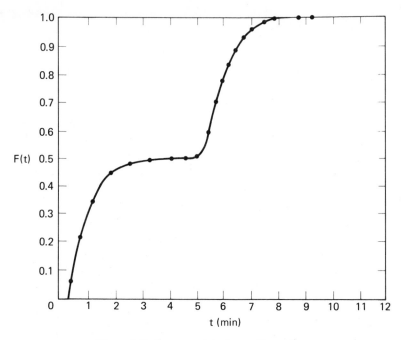

Figure 5.21 Response data for problem 11

Section 5.4a

12. The reactor described in problem 3, chapter 4, is to be used for the reaction
    $A + B \longrightarrow C + D$, with order corresponding to stoichiometry. Predict the con-
    version at the outlet under isothermal conditions and compare with that for a
    PFR of the same average residence time. The following parameters apply:

$$C_{A_0} = 0.01211\ M$$

$$C_{B_0} = 0.02578\ M$$

$$C_{C_0} = C_{D_0} = 0$$

$$k = 0.57\ \text{liter/mole-sec}$$

$$\text{length} = 15.3\ \text{cm}$$

13. A tubular flow reactor exhibits a residence-time distribution which can be
    modeled by a sequence of 10 CSTR in series (all of the same volume). The
    nominal residence time in the tubular reactor is 20 sec. Compare, for a Type III
    reaction system, the conversion, selectivity, and yield obtained in the reactor
    (as modeled by the CSTR sequence) with that which would be obtained in a
    true PFR of the same residence time. The rate constants are, $k_1 = 0.1\ \text{sec}^{-1}$ and
    $k_2 = 0.05\ \text{sec}^{-1}$.

14. In Figure 5.12a the effect of nonideal flow on conversion in a tubular flow reactor was presented in terms of the CSTR model for a first-order reaction. Repeat this calculation for a second-order reaction and a half-order reaction with the same numerical value of the rate constant and an inlet concentration of reactant of 1.0 M. Compare the magnitude of the effect on conversion among the three different types of kinetics.

15. Make a comparison of the selectivity behavior of a CSTR sequence and approach to PFR behavior, similar to that of Figure 5.12b, for a Type III system in which
   (a) The kinetics of the first reaction are half-order.
   (b) The kinetics of the first reaction are second-order.

16. What is the ratio between elements of residence time $(t/\bar{t}) \leq 1$ and $(t/\bar{t}) \geq 1$ in the CSTR model for $n = 5$, 10, and 25?

Section 5.4b

17. (a) Given the residence time distribution data in Figure 5.22, determine the corresponding Peclet number based on overall reactor length.

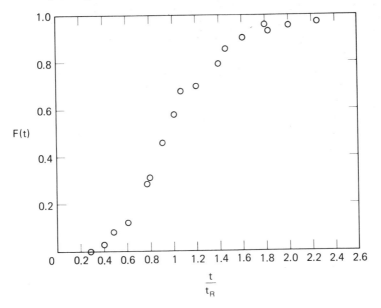

Figure 5.22 Residence-time distribution data for problem 17

   (b) Evaluate the conversion for a first-order reaction, rate constant $k = 0.0796$ sec^{-1}, in this reactor using the dispersion model. Residence time is 17.4 sec.

18. (a) Compare the conversion calculated in problem 17 with that determined

from the solution of Wehner and Wilhelm, for equal Peclet numbers in
the three sections.

(b) Repeat if $N_{Pe(a)} = N_{Pe(b)} = 0.5N_{Pe(c)}$.

19. Compare the conversion for a first-order reaction as predicted by the dispersion
model for Types A, B, and C boundary conditions. The parameters are $L =$
3.2 m, $u = 1.6$ m/sec, $k = 1.34$ sec^{-1}, and $N_{Pe} = 10$.

20. Consider the rate investigations of Sinfelt et al. in problem 12, chapter 4. If the
catalyst particle diameter is $\frac{1}{16}$ in. (spherical), what would be the influence of
axial dispersion on conversion for a 5% n-pentane/95% hydrogen feed at the
conditions indicated? A value for N_{Pe} may be determined from Figure 5.9b.

21. Evaluate the results obtained in problems 12 and 20, chapter 4, in terms of the
criterion developed by Mears for freedom from axial dispersion effects in tubular
reactors [D. E. Mears, *Ind. Eng. Chem. Proc. Design Devel.*, **10**(4), 541 (1971)].

22. The saponification of ethyl acetate in aqueous solution,

$$CH_3-C\overset{O}{\underset{OC_2H_5}{\diagdown}} + NaOH \longrightarrow CH_3-C\overset{O}{\underset{ONa}{\diagdown}} + C_2H_5OH$$

is a second-order, irreversible reaction whose rate constant at 25°C has been
measured as 8.0×10^{-2} liter/mole-sec [K. J. Laidler and D. Chen, *Trans.
Faraday Soc.*, **54**, 1026 (1958)]. A tubular reactor, 3 in. in length and 1 cm in
radius is available. Determine the total amount of feed (in. cm^3/sec) which can
be processed for

(a) 60% conversion at exit

(b) 98% conversion at exit

using 1 M concentrations of each reactant at the inlet. Use a PFR model first,
then repeat for a one-dimensional dispersion model. (N_{Pe} may be determined
from Figure 5.8b in the latter case.) Is the use of PFR or dispersion models consistent with the required flow rates?

(c) The activation energy is 15 kcal/mole. Repeat the calculation above for
$T = 85°C$. Neglect density changes with temperature.

23. A certain first-order reaction is being carried out in a fixed-bed reactor in which
the axial dispersion coefficient has been determined to be 3×10^{-3} cm^2/sec.
The linear velocity within the reactor is 5 cm/sec and the rate constant is
0.25 sec^{-1}.

(a) How long should the reactor be in order to have negligible effects of dispersion on the conversion? (Judgment is required here.)

(b) What is the conversion?

Section 5.4c

24. Derive general expressions for computing conversion for the second-order reaction $2A \longrightarrow B + C$ for the two combined models of problems 2 and 13, chapter 4.

25. Derive an equation for the conversion in a first-order irreversible reaction using
the combined model you proposed to fit the $E(t)$ data in problem 11.

Section 5.4d

26. A reaction that has been shown to be first-order irreversible with $k = 0.8$ hr^{-1} at 250°C is to be carried out in three CSTR with $\bar{t} = 1$ hr. Compare these two techniques of starting
   (a) Charging to each tank the raw feed at 250°C, running them batchwise until proper conversions are reached, and then starting flow.
   (b) Charging to each tank the raw feed at 250°C, and starting flow immediately. Which would you recommend, and why?
27. Consider the solution for the transient response of the dispersion model to a step-function input in concentration given in Figure 5.20 for the particular case of $N_{Pe} = 10$ and $R' = 0.1$. Is the transient response of a comparable CSTR series model the same?

NOTATION FOR CHAPTER 5

Section 5.2

$C(t), E(t), F(t)$	age distribution functions
θ	number of residence times

Section 5.3

A	cross-sectional area
C_0, C_{t_0}	inlet concentrations, moles/volume
C_{0^+}, C_{0^-}	concentrations at $Z = 0^+$ and 0^-, moles/volume
C_L, C_{out}	exit concentrations, moles/volume
C_n	concentration in nth vessel, moles/volume
D_M	molecular diffusion coefficient, area/time
D, D_e	axial dispersion coefficient, area/time
D_r	radial dispersion coefficient, area/time
d, d_p	reactor or particle diameter, length
$E_1(x)$	exponential integral
erf (y)	error function integral
f	fractional recirculation between stages
k	rate constant, units depend on order
L	reactor length, length
n	number in CSTR sequence
N_{Pe}	Peclet number, Lu/D
N_{Pe_r}	radial Peclet number, $\bar{u}d_p/D_r$
N_{Re}	Reynolds number, $d\bar{u}\rho/\mu$

N_{Sc}	Schmidt number, $\mu/\rho D_m$
r	radial coordinate, length
R	reactor radius, length
$t, \bar{t}, t_R$	time, time per vessel (CSTR), residence time (PFR)
$\bar{u}, u_0$	average and superficial velocity, length/time
$u, u(r)$	velocity or velocity at r, length/time
$\bar{V}, \bar{V}_n$	vessel volume
$x(t), \bar{x}$	conversion and average conversion
z	axial coordinate, length
ϵ	void fraction
η	geometric constant, equation (5-47)
μ	viscosity, mass/length-time
ν	$z - ut$
ρ	density, mass/volume
τ	tortuosity

Section 5.4

f	fraction active volume in combined model
v, v_1, v_2	volumetric flow rates, volume/time
$\delta(t_1)$	defined as $\delta(t_1) = 0$ for $t < t_1$
	$\delta(t_1) = 1$ for $t > t_1$

Section 5.5

A_n	concentration of A in n at time zero, moles/volume
f	dimensionless concentration, C/C_0
k_1, k_2	rate constants for two steps of Type II or III, time^{-1}
n	reaction order
Q	$1 + k\bar{t}$
R', R'_n	(kL/u), $(2kLn/u)C_0^{n-1}$, reaction parameters
$(-r)$	rate of reaction, moles/volume-time
$S_B(II,III)_{PFR}$	PFR selectivity for B, Type II or III
$S_B(II,III)_{CSTR}$	CSTR selectivity for B, Type II or III
$S_i(III)$	intrinsic selectivity, k_1/k_2, Type III
x_{PFR}	PFR conversion
α	$(1 + k\bar{t})$ or $(1 + k_1\bar{t})$, Type III
β	$(1 + k_2\bar{t})$, Type III, or $(1 + 4R'/N_{Pe})^{1/2}$, equations (5-71) to (5-85)
γ	$1 + (k_1 + k_2)\bar{t}$, Type II
ζ	dimensionless length

Thermal Effects **6**
in the Modeling
of Real Reactors

As we have seen in chapter 4, the analysis of nonisothermal reactors in any general way, even for the ideal PFR and CSTR, is not possible to the same degree of elegance as for isothermal reactors because of the intractable nonlinearity of the Arrhenius equation. When we graduate to the more complicated models for the nonideal systems we have been discussing, the problem is, of course, compounded. For example, while the CSTR-sequence model retains its algebraic structure, the energy balance requires solution of an equation implicit in temperature for each unit; stagewise computation is required and much of the convenience of that model is lost. Similarly, dispersion models require numerical solution. The result of this is a slight change of emphasis in this section with more attention given to actual procedures involved in solution of the nonisothermal models. Hopefully, the justifications for the choice of models for nonideal reactors have been established with sufficient care that we may profitably turn attention to these more practical matters.

In some of the examples immediately to follow, the phenomena of steady-state multiplicity or of parametric sensitivity will be encountered. At this point we will pass these along without extensive comment, with the intention of returning to the topic in some more detail at the end of the chapter.

6.1 Mixing-Cell Sequences

The mass and energy balances for an individual unit in the mixing cell sequence may be written in general form as

$$v(C - C_0) + \bar{V}(-r) = 0 \tag{6-1}$$

and

$$v(\theta_0 - \theta) + \bar{V}(-r) - \frac{U'}{C_p}(\theta - \theta_M) = 0 \tag{6-2}$$

where θ is a reduced temperature, $C_p T/(-\Delta H)$, U' is a total heat-transfer coefficient ($U' = UA$), and C_p is on a volumetric basis. When written in this form, the similarity of the two equations in the limit of adiabatic operation is apparent. The remainder of the nomenclature here is as defined in Figure 4.21. Now the sum of equations (6-1) and (6-2) may be written as

$$v(C_0 - C) + v(\theta_0 - \theta) - \frac{U'}{C_p}(\theta - \theta_M) = 0 \tag{6-3}$$

or

$$C = C_0 + (\theta_0 - \theta) + \frac{U'}{v C_p}(\theta_M - \theta) \tag{6-4}$$

If we specify the concentration in the unit, the reaction rate is fixed according to

$$(-r) = (C_0 - C)\frac{v}{\bar{V}} = \frac{C_0 - C}{\bar{t}} \tag{6-5}$$

and the temperature is fixed by the form of temperature dependence of $(-r)$. For a first-order irreversible reaction, thus,

$$T = \frac{E/R}{\ell n\,[k_0 \bar{t}\, C/(C_0 - C)]} \tag{6-6}$$

Equations (6-4) to (6-6) may be used to establish an *operating diagram* for this CSTR unit [C. R. McGowin and D. D. Perlmutter, *Amer. Inst. Chem. Eng. J.*, **17**, 831 (1971)] of the type shown in Figure 6.1 for the case of first-order irreversible kinetics. The operating temperature is not shown explicitly on these curves and would have to be determined separately. The condition of steady-state multiplicity is indicated on the diagram, for example, by line *a*, at the three points of intersection with the operating curve for $U'/v C_p = 0.3$.

The operating diagram as established in this manner is a convenient way to display some overall picture of temperature effects on CSTR operation, but it is not going to take us very far in terms of modeling nonideal reactors via mixing-cell sequences because it is set up backward. That is, we specify conversion, ordinarily the quantity we are seeking, to define the conditions of operation. In fact, this coupling between the mass and energy

314

Parametric values:

$$k_o \bar{t} = 1.0 \times 10^{-11}$$

$$\frac{C_p (E/R)}{(-\Delta H) C_o} = 75$$

$$\frac{C_p T_M}{(-\Delta H) C_o} = 2.5$$

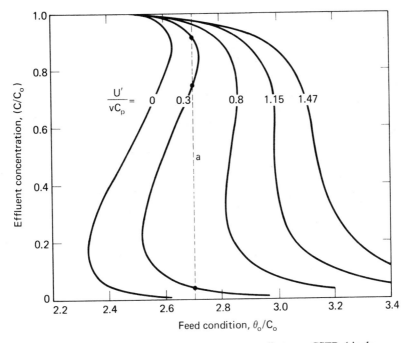

Figure 6.1 Operating diagram for temperature effects on CSTR (single-unit) operation (after D. D. Perlmutter, *Stability of Chemical Reactors,* ©️ 1972; reprinted by permission of Prentice-Hall, Inc., Englewood Cliffs, N.J.)

conservation relationship makes it rather inconvenient to use the CSTR sequence as a model for nonideal reactors in nonisothermal operation and, as a consequence, not too much attention has been given to this approach in the literature. The crux of the problem may be illustrated using equations (6-1) and (6-2) for the simplest case of adiabatic operation and irreversible first-order kinetics:

$$v(C_0 - C) - \bar{V}kC = 0 \qquad (6\text{-}1a)$$

$$v(\theta_0 - \theta) - \bar{V}kC = 0 \qquad (6\text{-}2a)$$

Solving for C from (6-1a) gives in the usual fashion:

$$C = \frac{C_0}{1 + k\bar{t}}$$

Replacing this in equation (6-2a) yields the general relationship to be satisfied by temperature and concentration:

$$\theta_0 - \theta - \frac{k\bar{t}C_0}{1 + k\bar{t}} = 0 \qquad (6-7)$$

There is no convenient, explicit solution for temperature available from this equation, so that, in large measure, the attractive features of the CSTR sequence model as used in the isothermal case have been lost.

Some further problems surface if we press the inquiry to more detail for the general nonisothermal situation. In the nonadiabatic case a heat-transfer term appears in equation (6-2) and one must be concerned with the question of how to determine a proper CSTR heat-transfer coefficient to provide a reasonable representation of the heat-transfer situation in the reactor being modeled. Also, in isothermal analysis we have proposed determining the number of cells to be used in the CSTR sequence via matching of residence time distribution curves. In terms of the dispersion model, this also implies a corresponding value of the axial dispersion coefficient or the Peclet number. If we follow the same procedure in nonisothermal cases (presumably using RTD information obtained under isothermal conditions) we are imposing implicitly a value for the axial dispersion of heat on the CSTR model which may or may not be valid. A better appreciation of this point will be provided the reader in the discussion of nonisothermal dispersion models.

In spite of the difficulties and inconveniences pointed out above, a very detailed and elegant mixing-cell model has been developed for nonisothermal reactors by Deans and Lapidus [H. A. Deans and L. Lapidus, *Amer. Inst. Chem. Eng. J.*, **6**, 656 (1960)], which is capable of handling radial as well as longitudinal gradients (i.e., two-dimensional). It is a computational model developed with numerical rather than analytical solutions in mind; although some aspects of this approach extend beyond the general scope of our present effort, we present this development in Appendix A to this chapter.

6.2 Axial Dispersion Models

The one-dimensional dispersion model, in contrast to stirred tank sequences, has received considerable attention in the literature in application to non-ideal reactors under nonisothermal conditions. Physical intuition tells us that when heat transfer exists at the wall, there must be temperature gradients in radial as well as axial dimensions (also pointed out in the mixing-cell model of Deans and Lapidus), so a first important question to be asked is how adequate and under what conditions does a one-dimensional model provide description of two-dimensional reality. This question has been addressed by a number of workers; most recent studies indicate that the one-dimensional axial dispersion model is capable of accounting at least for parabolic radial temperature profiles if the wall heat-transfer coefficient is chosen properly. This has been discussed in detail by Froment [G. F. Froment, *Ind. Eng. Chem.*, **59**, 18 (1967)] and more recently by Finlayson (B. A. Finlayson, *The Method of Weighted Residuals and Variational Principles*, Academic Press, New York, 1972). With this caveat in mind, we will focus attention in this section to the development of nonisothermal one-dimensional dispersion models.

The mass balance is, of course, unchanged from the isothermal case, equation (5-65), so we have

$$D\frac{d^2C}{dz^2} - u\frac{dC}{dz} - (-r) = 0 \qquad (5\text{-}65)$$

The corresponding energy balance may be derived by a procedure similar to that used in obtaining equation (5-65), but it is perhaps simpler to modify equation (4-121) previously derived for the nonisothermal PFR:

$$\rho u C_p \frac{dT}{dz} = (-r)(-\Delta H) - \frac{2U}{R}(T - T_M) \qquad (4\text{-}121)$$

Since we know that the net effect of the dispersion model is to add a diffusional term to the PFR model, equation (4-121) must become for the dispersion model:

$$\rho C_p E'\frac{d^2T}{dz^2} - \rho u C_p \frac{dT}{dz} + (-r)(-\Delta H) - \frac{2U}{R}(T - T_M) = 0 \qquad (6\text{-}8)$$

where E' is a transport coefficient for heat analogous to D for mass. As in the previous case, u, ρ, and C_p are taken to be constant with position and conversion, which in some instances may not be reasonable. The positive sign on the reaction-rate term is in accord with the convention on $(-\Delta H)$, which for an exothermic reaction is a positive number, and on $(-r)$, which is a positive quantity when with reference to reactants concentration.

For adiabatic reactors equations (5-65) and (6-8) may be treated by a somewhat simplified analysis, since the mass and energy balances can be uncoupled from each other, as shown by van Heerden [C. van Heerden, *Chem. Eng. Sci.*, **8**, 133 (1958)]. Let us define a special temperature variable, θ, as before:

$$\theta = \frac{C_p T}{(-\Delta H)} \qquad (6\text{-}9)$$

On substituting this into equation (6-8), we obtain

$$E'\frac{d^2\theta}{dz^2} - u\frac{d\theta}{dz} + (-r) = 0 \qquad (6\text{-}10)$$

where the heat-transfer term disappears because of adiabatic operation. Now the mass and energy balances equations (5-65) and (6-10) can be added with the elimination of the rate term to give

$$E' \frac{d^2\theta}{dz^2} + D \frac{d^2C}{dz^2} = u\left(\frac{d\theta}{dz} + \frac{dC}{dz}\right) \qquad (6\text{-}11)$$

If the mechanisms for mass and energy dispersion are the same (a topic of some debate, particularly for nonhomogeneous reactors), then $E' = D$ and

$$D\left(\frac{d^2\theta}{dz^2} + \frac{d^2C}{dz^2}\right) = u\left(\frac{d\theta}{dz} + \frac{dC}{dz}\right) \qquad (6\text{-}12)$$

Definition of a new variable,

$$\omega = \theta + C \qquad (6\text{-}13)$$

reduces equation (6-12) to

$$\frac{d^2\omega}{dz^2} - \frac{u}{D}\frac{d\omega}{dz} = 0 \qquad (6\text{-}14)$$

with boundary conditions as discussed previously:

$$\omega = \omega_0 - \frac{D}{u}\left(\frac{d\omega}{dz}\right) \qquad (z = 0)$$

$$\frac{d\omega}{dz} = 0 \qquad (z = L)$$

The solution to equation (6-14) under these conditions is simply

$$\omega = C_0 + \theta_0 = \text{constant} = C + \theta \qquad (6\text{-}15)$$

Thence,

$$C = C_0 + \theta_0 - \theta$$

The energy balance can now be written in terms of temperature alone, with $(-r)$ given by

$$(-r) = k_0 \exp\left[-EC_p/R\theta(-\Delta H)\right](C_0 + \theta_0 - \theta) \qquad (6\text{-}16)$$

Methods of solution for equations (6-10) and (6-16) are obviously numerical, since the simplified adiabatic model is still nonlinear. Solutions to this set of equations have been reported by Raymond and Amundson [L. R. Raymond and N. R. Amundson, *Can. J. Chem. Eng.*, **42**, 173 (1964)], who report the possibility of steady-state multiplicity for certain ranges of parameter values. We shall discuss this possibility, and the details of a solution procedure for this and other nonisothermal models later in this section.

For solution of the more general nonisothermal, nonadiabatic model, it is more convenient to write the mass and energy balances in terms of the conventional reduced variables for concentration and temperature:

$$f = \frac{C}{C_0} \qquad \tau = \frac{T}{T_0}$$

This leads to

$$\frac{1}{N_{\mathrm{Pe}_m}}\frac{d^2f}{d\zeta^2} - \frac{df}{d\zeta} - (-r)\left(\frac{L}{u}\right) = 0 \tag{6-17}$$

$$\frac{1}{N_{\mathrm{Pe}_h}}\frac{d^2\tau}{d\zeta^2} - \frac{d\tau}{d\zeta} + \frac{(-r)(-\Delta H)}{\rho C_p T_0}\left(\frac{L}{u}\right)C_0 - \frac{2U}{\rho C_p R}\left(\frac{L}{u}\right)(\tau - \tau_M) = 0 \tag{6-18}$$

with the boundary conditions

$$\frac{df(0^+)}{d\zeta} = N_{\mathrm{Pe}_m}[f(0^+) - 1] \qquad (\zeta = 0)$$

$$\frac{d\tau(0^+)}{d\zeta} = N_{\mathrm{Pe}_h}[\tau(0^+) - 1] \qquad (\zeta = 0) \tag{6-19}$$

$$\frac{d\tau}{d\zeta} = \frac{df}{d\zeta} = 0 \qquad (\zeta = 1)$$

In general, regardless of the mathematical form of $(-r)$, these equations will require numerical solution. We present here a scheme, for u, ρ, and C_p independent of position and conversion, patterned after the development of Carberry and Wendel [J. J. Carberry and M. M. Wendel, *Amer. Inst. Chem. Eng. J.*, **9**, 129 (1963)], which has proven to be of considerable utility in solution of diffusional-type boundary-value problems.

Consider the mass balance, equation (6-17). Let us introduce the following finite-difference forms for the derivatives:

$$\frac{d^2f}{d\zeta^2} = \frac{f_{p+1} - 2f_p + f_{p-1}}{\Delta\zeta^2}$$

$$\frac{df}{d\zeta} = \frac{f_{p+1} - f_p}{\Delta\zeta}$$

where p, $p-1$, and $p+1$ are points along an axial grid of N points separated by $\Delta\zeta$ from each other such that $N\,\Delta\zeta = 1$. Thus, equation (6-17) becomes

$$\frac{1}{N_{\mathrm{Pe}}} \cdot \frac{f_{p+1} - 2f_p + f_{p-1}}{\Delta\zeta^2} - \frac{f_{p+1} - f_p}{\Delta\zeta} - (-r)_p\left(\frac{L}{u}\right) = 0 \tag{6-20}$$

where subscript m has been dropped from N_{Pe}. If we include our familiar first-order kinetics as an example, then

$$(-r)_p = kf_p$$

and equation (6-20) may be written

$$\frac{1}{N_{\mathrm{Pe}}}f_{p-1} + \left(\Delta\zeta - \frac{2}{N_{\mathrm{Pe}}} - R'\,\Delta\zeta^2\right)f_p + \left(\frac{1}{N_{\mathrm{Pe}}} - \Delta\zeta\right)f_{p+1} = 0 \tag{6-21}$$

with

$$R' = k_p\frac{L}{u} = \frac{L}{u}k_0 e^{-\gamma/\tau_p}$$

$$\gamma = \frac{E}{RT_0}$$

Now let us put the reaction-rate terms of equation (6-21) on the right-hand side such that, after some rearrangement, we obtain

$$f_{p-1} + (N_{Pe}\,\Delta\zeta - 2)f_p + (1 - N_{Pe}\,\Delta\zeta)f_{p+1} = N_{Pe}\left(\frac{L}{u}\right)\Delta\zeta^2 k_0 e^{-\gamma/\tau_p}f_p \quad (6\text{-}22)$$

This equation is of the form

$$f_{p-1}A_p + f_p B_p + f_{p+1}C_p = D_p \quad (6\text{-}23)$$

for point p in the axial grid. The coefficients are

$$A_p = 1$$
$$B_p = N_{Pe}\,\Delta\zeta - 2$$
$$C_p = 1 - N_{Pe}\,\Delta\zeta \quad (6\text{-}23a)$$
$$D_p = N_{Pe}\left(\frac{L}{u}\right)\Delta\zeta^2 k_0 e^{-\gamma/\tau_p}f_p$$

For the whole reactor, then, with the exception of the initial point, $p = 1$, and the final point, $p = N$, we have the following set of equations:

$$f_1 A_2 + f_2 B_2 + f_3 C_2 = D_2$$
$$f_2 A_3 + f_3 B_3 + f_4 C_3 = D_3$$
$$\cdot$$
$$\cdot$$
$$\cdot$$
$$f_{p-1}A_p + f_p B_p + f_{p+1}C_p = D_p \quad (6\text{-}24)$$
$$\cdot$$
$$\cdot$$
$$\cdot$$
$$f_{N-2}A_{N-1} + f_{N-1}B_{N-1} + f_N C_{N-1} = D_{N-1}$$

At the inlet we may write

$$f_0 A_1 + f_1 B_1 + f_2 C_1 = D_1 \quad (6\text{-}25)$$

but the boundary condition requires that

$$f_0 = 1 + \frac{1}{N_{Pe}}\left(\frac{df}{d\zeta}\right)_0 \quad (6\text{-}26)$$

Initial gradients are best estimated by higher-order difference formulas such as

$$\left(\frac{df}{d\zeta}\right)_0 = \frac{-3f_0 + 4f_1 - f_2}{2\Delta\zeta} \quad (6\text{-}27)$$

Substituting equations (6-26) and (6-27) for f_0 in equation (6-25) and rearranging,

$$f_1 B_1 + f_2 C_1 = D_1 - A_1\left[\frac{1}{N_{Pe}}\left(1 + \frac{-3f_0 + 4f_1 - f_2}{2\Delta\zeta}\right)\right]$$

or

$$f_1 B_1 + f_2 C_1 = D_1' \quad (6\text{-}25a)$$

At the exit we have

$$f_{N-1}A_N + f_N B_N + f_{N+1}C_N = D_N \qquad (6\text{-}28)$$

and the boundary condition requires that

$$\frac{f_{N+1} - f_N}{\Delta\zeta} = 0$$

so

$$f_{N+1} = f_N$$

Substituting this into equation (6-28) gives

$$f_{N-1}A_N + f_N(B_N + C_N) = D_N$$

or $\qquad\qquad\qquad\qquad\qquad\qquad\qquad\qquad\qquad$ (6-28a)

$$f_{N-1}A_N + f_N B'_N = D_N$$

The full set of finite-difference equations for the mass balance on the reactor consists of equations (6-24) plus (6-25a) and (6-28a)

$$f_1 B_1 + f_2 C_1 = D'_1$$

$$\cdot$$
$$\cdot$$
$$\cdot$$

$$f_{p-1}A_p + f_p B_p + f_{p+1}C_p = D_p \qquad (6\text{-}29)$$

$$\cdot$$
$$\cdot$$
$$\cdot$$

$$f_{N-1}A_N + f_N B'_N = D_N$$

In matrix form these may be written as

$$
\begin{bmatrix}
B_1 & C_1 & & & & \\
A_2 & B_2 & C_2 & & & \\
& A_3 & B_3 & C_3 & & \\
& & \cdot & \cdot & \cdot & \\
& & & \cdot & \cdot & C_{N-1} \\
& & & & A_N & B'_N
\end{bmatrix}
\begin{bmatrix}
f_1 \\ f_2 \\ f_3 \\ \cdot \\ \cdot \\ \cdot \\ f_{N-1} \\ f_N
\end{bmatrix}
=
\begin{bmatrix}
D'_1 \\ D_2 \\ D_3 \\ \cdot \\ \cdot \\ \cdot \\ D_{N-1} \\ D_N
\end{bmatrix}
\qquad (6\text{-}30)
$$

The form of the matrix given in equation (6-30) provides justification for some of the manipulations employed in the finite-difference equations, which may have seemed to be rather arbitrary. The matrix of equation (6-30) is termed "tridiagonal," for there are no nonzero elements off the three diagonal entries indicated. For matrices of this type there exists a simple and direct method of inversion, and thus solution of the set of equations, known as the *Thomas method*. The computational procedure is as follows. Determine

$$w_1 = B_1$$

$$w_r = B_r - A_r q_{r-1} \qquad [r = 2, 3, \ldots, N(\text{``}B\text{''}_N = B'_N)]$$

where

$$q_{r-1} = \frac{C_{r-1}}{w_{r-1}} \tag{6-31}$$

Also,

$$g_1 = \frac{D'_1}{w_1} = \frac{D'_1}{B_1}$$

$$g_r = \frac{D_r - A_r g_{r-1}}{w_r} \qquad (r = 2, 3, \ldots, N) \tag{6-32}$$

The solution to the set of equations is given as

$$f_N = g_N$$
$$f_r = g_r - q_r f_{r+1} \qquad r = 1, 2, \ldots, N-1 \tag{6-33}$$

According to this algorithm, then, we calculate values of w and g for increasing values of the index r, starting at 1, then compute values of f starting at f_N and going back to f_1. Further details of the Thomas method are given by Lapidus (L. Lapidus, *Digital Computation for Chemical Engineers*, McGraw-Hill, New York, 1962, p. 254).

The way these equations have been set up, the values of the coefficients D_p depend on the values of f_p and τ_p; that is, on what the temperature and concentration profiles are. In addition, equation (6-30) is for the material balance only. To solve the nonisothermal reactor, it is necessary to construct a corresponding tridiagonal system for the energy balance and solve the two sets of equations simultaneously. The same procedures may be used to construct the energy equations as detailed above for the mass balance; coefficients in the resultant matrix will also in general be dependent on f_p and τ_p. It is in this aspect that the simplicity of the Thomas method algorithm is of great importance, since a trial-and-error procedure may be employed without excessive computer-time requirements. The trial procedure is simple; since we are dealing with reduced variables, one merely starts the computation with τ_p and f_p equal to unity[1] at each grid point, solves the resultant matrices via equations (6-31) to (6-33), uses the resulting sets of τ_p and f_p to determine new values for the matrix elements, and iterates on this. Final solution is indicated when the variation between τ and f profiles from one iteration to the next is less than some specified minimum. Selection of an appropriate minimum ϵ may require some computational experiments, since there can be wide variation from case to case. The following types of con-

[1]Convergence may be assisted by setting τ_p and f_p equal to unity only at the reactor entrance and at different values elsewhere to assist in numerical evaluation of the derivative in the inlet boundary condition.

vergence criteria have been proposed:

$$\sum_{p=1}^{N} (\tau_{i-1} - \tau_i)_p \le \epsilon_1$$

$$\sum_{p=1}^{N} (f_{i-1} - f_i)_p \le \epsilon_1 \tag{6-34}$$

or

$$\text{Max } (\tau_{i-1} - \tau_i) \le \epsilon_2$$

$$\text{Max } (f_{i-1} - f_i) \le \epsilon_2 \tag{6-35}$$

where $i - 1$ and i are the next-to-last and last iteration steps, respectively. Equations (6-34) propose that the sum of the deviations on both profiles in the last two iterations be $\le \epsilon_1$, while equations (6-35) propose that the maximum differences in iteration for the two profiles be $\le \epsilon_2$. For comparable conditions of convergence ϵ_2 will be roughly 0.01 to 0.1 ϵ_1, dependent on the number of grid points. Both types of convergence criteria have been used successfully in the literature.

The particular form of the coefficients D_p indicated in equation (6-23a) depends on the form of kinetics under consideration. A more general derivation not restricted to the first-order case would carry the corresponding expression for R' into the right-hand term, giving a different form for D_p. For power-law kinetics, for example, an appropriate nondimensional form of equation (5-86) may be used. With this modification to the coefficients, the procedure of equations (6-31) to (6-33) may be used to obtain numerically the solutions previously discussed for the isothermal one-dimensional dispersion model (cf. Figure 5.15).

We have used the tridiagonalization and Thomas method procedure here as an example because it is relatively straightforward and is easily rendered into a workable computer program. The reader should not, however, think that this is the only way in which systems of equations such as (5-65) and (6-8) may be solved and, indeed, the method here described may encounter difficulty in convergence for certain ranges of parameter values, particularly for endothermic reactions. In such cases redefinition of the coefficients may help, as well as exploration of the convergence critera employed. Alternative approaches for numerical solution are contained in the works of Douglas [J. Douglas, Jr., "A Survey of Numerical Methods for Parabolic Differential Equations," *Advan. Computers*, **2** (1961)], Von Rosenberg (D. U. Von Rosenberg, *Methods for the Numerical Solution of Partial Differential Equations*, American Elsevier, New York, 1969), and Finlayson (B. A. Finlayson, *The Method of Weighted Residuals and Variational Principles*, Academic Press, New York, 1972), among others.

Some typical solutions for the nonisothermal one-dimensional model, again in the form of operating diagrams, are given in Figure 6.2 for para-

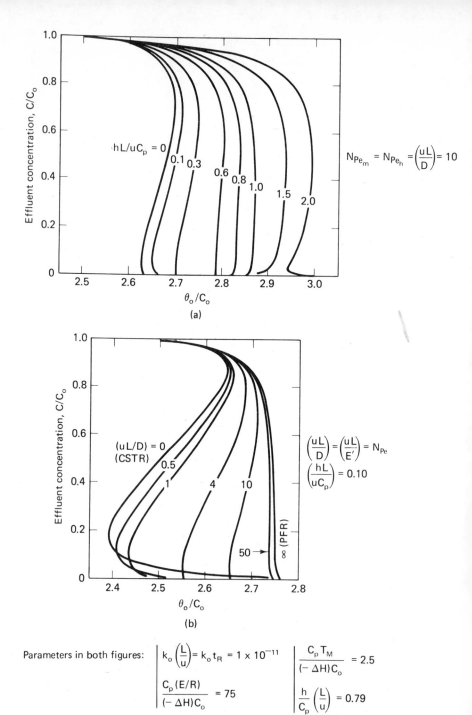

Figure 6.2 Steady-state operating diagrams for one-dimensional axial dispersion model (after D. D. Perlmutter, *Stability of Chemical Reactors*, © 1972; reprinted by permission of Prentice-Hall, Inc., Englewood Cliffs, N.J.)

where:

$$a_{11} = \frac{v}{V} + k_0 e^{-E/RT_s}$$

$$a_{12} = k_0 C_{A_s} \left(\frac{E}{RT_s^2}\right) e^{-E/RT_s}$$

$$a_{21} = k_0 \frac{(-\Delta H)}{\rho C_p} e^{-E/RT_s}$$

$$a_{22} = \frac{v}{V} + \frac{k_0(-\Delta H)}{\rho C_p} C_{A_s} \left(\frac{E}{RT_s^2}\right) e^{-E/RT_s} + \frac{UA}{\rho C_p V}$$

It is not necessary to solve equations (6-38) and (6-39) analytically for the stability analysis, although this can be done. If the system is stable, the small perturbations a and b must approach zero as time increases. Now, in matrix notation we may write these equations as

$$\frac{d\mathbf{x}}{dt} = \mathbf{A}\mathbf{x} \tag{6-40}$$

with the coefficient matrix A given by

$$\mathbf{A} = \begin{bmatrix} -a_{11} & -a_{12} \\ -a_{21} & -a_{22} \end{bmatrix}$$

We know from matrix theory that the solution to equation (6-40) is given by a sum of exponential terms in time in which the coefficients of time in the exponentials are the eigenvalues, λ, of the problem. Stability may thus be expected only if these eigenvalues contain negative real parts, so the exponential terms will die out as time increases. The values of λ are obtained from the characteristic equation:

$$\det \begin{bmatrix} -a_{11} - \lambda & -a_{12} \\ -a_{21} & -a_{22} - \lambda \end{bmatrix} = 0 \tag{6-41}$$

Expanding the determinant,

$$\lambda^2 + (a_{11} + a_{22})\lambda + a_{11}a_{22} - a_{12}a_{21} = 0 \tag{6-42}$$

or

$$\lambda_1, \lambda_2 = \frac{-(a_{11} + a_{22}) \pm [(a_{11} + a_{22})^2 - 4(a_{11}a_{22} - a_{12}a_{21})]^{1/2}}{2} \tag{6-43}$$

From examination of equation (6-43) it can be determined that the following conditions are required for the eigenvalues to have negative real parts:

$$a_{11} + a_{22} > 0$$

$$a_{11}a_{22} - a_{12}a_{21} > 0$$

or

$$\frac{2v}{\bar{V}} + k_0 e^{-E/RT_s} + \frac{UA}{\rho C_p \bar{V}} > k_0 e^{-E/RT_s} \frac{(-\Delta H)}{\rho C_p} C_{A_s} \left(\frac{E}{RT_s^2}\right)$$

$$\frac{v}{\bar{V}} + k_0 e^{-E/RT_s} + \frac{UA}{\rho C_p \bar{V}} + \frac{UAk_0 e^{-E/RT_s}}{v \rho C_p} > k_0 e^{-E/RT_s} \left(\frac{-\Delta H}{\rho C_p}\right) C_{A_s} \left(\frac{E}{RT_s^2}\right)$$

$$(6\text{-}44)$$

It is seen that these criteria for stability are given in terms of steady-state values; no transients need be calculated. Since the problem may be formulated in terms of a linear matrix, it is easily generalized to chemical reactions involving n species, either in simultaneous or sequential kinetic schemes. Further details of this are discussed in the original paper by Bilous and Amundson, and a clear description of the matrix methods involved is provided in the text by Perlmutter.

The stability criteria of equations (6-44) refer to small perturbations, a necessity of the Taylor series expansion. For large perturbations (the startup problem) we are interested in the approach to the steady state and, in cases where multiplicity exists, which steady state is approached. A qualitative, yet informative approach to this problem can be constructed as follows. Rewrite the balance equations in the following form:

$$\frac{dC_A}{dt} = A_1 - B_1 C_A - D_1 C_A e^{-E/RT} \qquad (6\text{-}45)$$

$$\frac{dT}{dt} = A_2 - B_2 T + D_2 C_A e^{-E/RT} \qquad (6\text{-}46)$$

where A_i, B_i, and D_i are constants > 0. When multiplicity exists, two lines may be plotted in the C_A–T plane corresponding to $(dC_A/dt) = 0$ and $(dT/dt) = 0$, lines 1 and 2, respectively, as shown qualitatively in Figure 6.4. Now, at a fixed value of T as C_A increases, the following must be true:

1. $A_1 - B_1 C_A - D_1 C_A e^{-E/RT}$ decreases
2. $A_2 - B_2 T + D_2 C_A e^{-E/RT}$ increases

Thus,

Above curve 1: $\dfrac{dC_A}{dt} < 0$

Below curve 1: $\dfrac{dC_A}{dt} > 0$

Above curve 2: $\dfrac{dT}{dt} > 0$

Below curve 2: $\dfrac{dT}{dt} < 0$

The phase-plane diagram of Figure 6.4 is divided into regions I to VI,

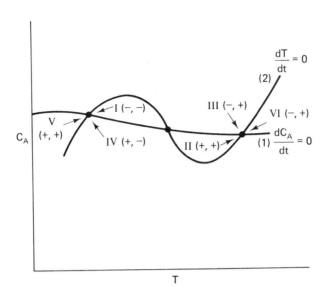

Figure 6.4 Phase-plane diagram (C_A–T) for a CSTR exhibiting steady-state multiplicity

each corresponding to a particular combination of the signs on the concentration and temperature derivatives. In region I, for example, according to the above reasoning $(dC_A/dt) = -$ and $(dT/dt) = -$, indicated as I $(-, -)$. Corresponding signs of the derivatives in the other sections are indicated on the figure. Each combination of signs indicates the direction of motion of the concentration/temperature behavior of the reactor in the unsteady state as shown by the arrows. In the phase plane the intersections of curves 1 and 2 determine potential steady-state operating points $(dC_A/dt = dT/dt = 0)$; it is seen that regardless of the precise location of a (C_A, T) coordinate representing a startup condition in the six regions, only the two outer steady states are approached. Whether the upper or lower state is that attained in steady operation is strictly a function of the signs of the concentration and temperature derivatives. The middle steady state is seen to be an unstable one; the reactor will not spontaneously move to this condition on startup.

The unstable nature of the intermediate steady state can also be envisioned in terms of the response to temperature perturbations of the heat-generation and heat-removal terms, q_g and q_r, defined for the CSTR in chapter 4 and illustrated in Figure 6.5 for steady-state multiplicity corresponding to the phase diagram of Figure 6.4. Consider the intersection of q_g and q_r at point A in Figure 6.5. As the temperature is increased, a small amount above the steady-state operating condition, T_A, the rate of heat removal, q_r, increases more rapidly than the rate of heat generation, q_g. Hence, the natural tendency of the system in response to this perturbation

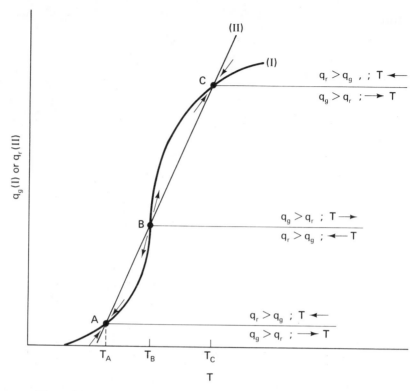

Figure 6.5 Heat generation/removal plot for a case of CSTR steady-state multiplicity

is to return to steady-state condition, T_A. Conversely, as the temperature is decreased a small amount below T_A, q_g becomes greater than q_r and the system will also return to T_A. Precisely the same situation pertains around the upper steady state at point C. For the midpoint B, however, the situation is reversed. In the event of a positive perturbation about T_B, q_g becomes greater than q_r and the system will move spontaneously to a point where heat-generation and heat-removal rates are in balance (i.e., C). Similarly, for negative perturbations about T_B, $q_r > q_g$ and the system will move to A.

It is of obvious interest to see if we can define in terms of the kinetic and reactor parameters, a criterion which would ensure the uniqueness of the steady-state operation of a CSTR. This can be done in a number of ways; we illustrate a simple one here. Start with the sum of the steady-state balances given by equation (6-4):

$$C = C_0 + (\theta_0 - \theta) + \frac{U'}{vC_p}(\theta_M - \theta) \qquad (6\text{-}4)$$

Substitution of this relationship into equation (6-2), the heat balance, gives

$$f(\theta) = 0 = v(\theta_0 - \theta) + \bar{V}[-r(\theta)] - \frac{U'}{C_p}(\theta - \theta_M) \tag{6-47}$$

where $[-r(\theta)]$ is now a function of the temperature variable θ only. Now in order for there to be only a single solution to equation (6-47) the function $f(\theta)$ can be zero at only one point. Since $f(\theta)$ is a continuous function, if it is zero at more than one point, the mean value theorem insists that $df(\theta)/d\theta$ must be zero at some point. Hence, if we can define conditions which ensure that this derivative will never be zero, uniqueness is assured. Differentiating equation (6-47) yields

$$\frac{df(\theta)}{d\theta} = -v + \bar{V}\frac{d(-r)}{d\theta} - \frac{U'}{C_p} \tag{6-48}$$

From this expression it is apparent that the derivative cannot be zero if $d(-r)/d\theta \leq 0$, or since $\theta = C_p T/(-\Delta H)$, if

$$\frac{d(-r)}{dT} \leq 0 \tag{6-49}$$

The specific form of the uniqueness criterion thus depends on the kinetics of the reaction. For nth-order irreversible kinetics,

$$(-r) = k_0 C^n e^{-E/RT}$$

then

$$\frac{d(-r)}{dT} = (-r)\left(\frac{E}{RT^2} + \frac{n}{C}\frac{dC}{dT}\right) \tag{6-50}$$

From equation (6-50) we obtain

$$\frac{d(-r)}{dT} \leq 0 \quad \text{if} \quad \frac{E}{RT^2} + \frac{n}{C}\frac{dC}{dT} \leq 0 \tag{6-51}$$

The concentration derivative (dC/dT) may be obtained from equation (6-4) as

$$\frac{dC}{dT} = -\frac{C_p}{(-\Delta H)}\left(1 + \frac{U'}{vC_p}\right) \tag{6-52}$$

Substituting equation (6-52) into (6-51) gives

$$\frac{E(-\Delta H)C}{nRC_pT^2} \leq 1 + \frac{U'}{vC_p}$$

In terms of inlet feed conditions, C_0 and T_0, the uniqueness criterion is

$$\frac{E(-\Delta H)C_0}{T_0^2 nRC_p} \leq 1 + \frac{U'}{vC_p} \tag{6-53}[1]$$

In fact, the criterion of equation (6-53) is quite a conservative one because

---

[1] Treatment of the inequalities here is based on the assumption of exothermic reaction with $T_0 > T_M$. If $T_M > T_0$, T_M^2 should be substituted for T_0^2 in equation (6-53).

of some approximations involved in defining the conditions required for $df(\theta)/d\theta \leq 0$. The use of $d(-r)/dT \leq 0$ will indeed ensure this, but a more precise definition of the required inequality from equation (6-48) would state that

$$\frac{d(-r)}{d\theta} < \frac{v}{\bar{V}} + \frac{U'}{\bar{V}C_p}$$

or, in terms of temperature,

$$\frac{d(-r)}{dT} < \alpha\frac{v}{\bar{V}} \tag{6-54}$$

where

$$\alpha = \frac{C_p}{(-\Delta H)} + \frac{U'}{v(-\Delta H)}$$

In the language of those concerned with such criteria $(d(-r)/dT) \leq 0$ forms a sufficient but not a necessary condition for steady-state uniqueness. The criteria that may be derived from the condition of equation (6-54) depend on whether the temperature range involved in the reactor operation includes the maximum of $d(-r)/dT$, as defined by $d^2(-r)/dT^2 = 0$, or whether $d(-r)/dT$ is monotonic in the region of reactor operation. In the former case,

$$T_0 \quad \text{or} \quad T_M \leq \frac{E/R}{2 + \psi}$$

and uniqueness is established by (for first-order kinetics)

$$\frac{v}{k_0\bar{V}} > \frac{\psi + 4}{\psi} \exp(-\psi - 2) \tag{6-55}$$

In the latter case,

$$T_0 \quad \text{and} \quad T_M \geq \frac{E/R}{2 + \psi}$$

and uniqueness is established by

$$\frac{v}{k\bar{V}} > \left[\frac{E}{RT_0^2}\left(\frac{E}{R\psi} - T_0\right) - 1\right]\exp\left(\frac{E}{RT_0}\right) \tag{6-56}$$

where

$$\psi = \frac{\alpha E/R}{C_0 + C_pT_0/(-\Delta H) + U'T/v(-\Delta H)}$$

Analysis of steady-state multiplicity in chemical reactors has been a topic of extensive research in recent years with a by-now considerable literature. For further samplers into this fascinating area the reader is referred to the text of Perlmutter, reviews by Aris and Schmitz, [R. Aris, *Chem. Eng. Sci.*, **24**, 149 (1969); R. A. Schmitz, *Advan. Chem.*, **148**, 156 (1975)], and the early work of Liu and Amundson on corresponding problems in fixed-bed reactors [S-L. Liu and N. R. Amundson, *Ind. Eng. Chem. Fundls.*, **1**, 200 (1962); **2**, 183 (1963); S-L. Liu, R. Aris, and N. R. Amundson, *Ind. Eng. Chem. Fundls.*, **2**, 12 (1963)].

b. Parametric sensitivity of a tubular reactor

The illustration of nonisothermal PFR behavior at the end of chapter 4, in Figure 4.20, showed extreme responses of temperature and partial pressure profiles to small changes in operating conditions in certain regions. Similarly, small variations in reactor parameters may result in such changes for fixed operating conditions. Such behavior has been termed *parametric sensitivity* of the reactor; the phenomenon is distinctly different from reactor stability in the sense that we have discussed that topic here, although the terms are occasionally used interchangeably in the literature.

The basic model which has been used in most studies of parametric sensitivity is the nonisothermal PFR for which

$$-u\frac{dC}{dz} = (-r) \tag{4-39}$$

$$\rho u C_p \frac{dT}{dz} = (-r)(-\Delta H) - \frac{2U}{R}(T - T_M) \tag{4-121}$$

with initial conditions C_0 and T_0 specified at the inlet, and with constant heat-transfer coefficient, U, and wall temperature, T_M.[1]

Obviously, the most important practical problem associated with parametric sensitivity is to determine conditions under which this condition is to be expected and what criteria may be established to ensure its absence. In 1959, Barkelew [C. H. Barkelew, *Chem. Eng. Progr. Symp. Ser.*, **55**(25), 37 (1959)] reported the results of a study aimed at determining regions of operating conditions where parametric sensitivity might exist. The analysis was based on the systematic numerical integration of equations (4-39) and (4-121) for many, many sets of parameters and the results were presented in the form of correlations in terms of kinetic and operating parameters, as illustrated in Figure 6.6, a composite of Barkelew's results as presented in the text by Perlmutter. For each type of kinetics the region in the figure below the solid line indicates a region where parametric sensitivity may exist. In addition, the region surrounding the solid line for the first-order case indicates the magnitude of variation in the region of parametric sensitivity as the inlet temperature is allowed to vary from the wall temperature within the range $T_M \pm (T_M^2 R/E)$. The regions indicated for zero-order and second-order kinetics pertain to $T_0 = T_M$. The regions of parametric sensitivity indicated on the Barkelew diagram indicate only potential problems (where the "hot-spot"

[1]In fact, this corresponds relatively well to an important class of reactors in industrial practice which are subject to parametric sensitivity. Hydrocarbon oxidation reactions, highly exothermic, are often carried out in reactors resembling large shell-and-tube heat exchangers, the tube side containing catalyst and the shell side some heat-transfer medium, such as molten salt, which serves to maintain wall temperature essentially constant with reactor length.

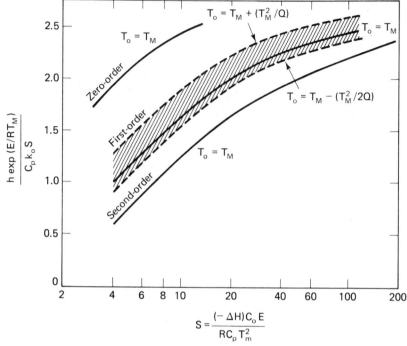

Figure 6.6 Regions of parametric sensitivity in nonisothermal PFR operation (after D. D. Perlmutter, *Stability of Chemical Reactors*, © 1972; reprinted by permission of Prentice-Hall, Inc., Englewood Cliffs, N.J.)

temperature may vary greatly with system parameters); the magnitude of this sensitivity or its relationship to individual parameters is not indicated and would have to be evaluated on an individual case-by-case basis. As such, the criteria established by diagrams such as Figure 6.6 tend to be relatively conservative, and alternative approaches have been developed to produce more precise a priori criteria defining parametric sensitivity.

To approach the problem more quantitatively, let us focus attention on the *hot-spot temperature*, that is, the maximum of the temperature profile along the length of the reactor. At this point, $(dT/dz) = 0$, so the rate of heat generation is equal to the rate of heat removal. Hence, from equation (4-121),

$$(-r)(-\Delta H) = \frac{2U}{R}(T_{\max} - T_M) \qquad (6\text{-}57)$$

where $T_{\max}$ is the maximum temperature along the reactor length. The left-hand side of equation (6-57) is a heat-generation term and the right-hand side a heat-removal term. We can thus establish some control over sensitivity by specification of some allowable T_{max} and manipulation of the heat-transfer coefficient and/or wall temperature such that

$$[-r(T_{max}, C_{max})] \leq \frac{2U}{(-\Delta H)R}(T_{max} - T_M) \qquad (6\text{-}58)$$

where $[-r(T_{max}, C_{max})]$ is the rate of reaction at the temperature maximum and C_{max} is the reactant concentration at this point. In fact, equation (6-58) is still not very much help, since if we are to know the rate at (T_{max}, C_{max}) we must solve the full problem. If we are willing, however, to accept a degree of conservatism in this analysis, we may note that

$$[-r(T_{max}, C_{max})] < [-r(T_{max}, C_0)]$$

where C_0 is the reactor inlet concentration. If T_{max} is specified, then $[-r(T_{max}, C_0)]$ is known and we may use as a design basis:

$$[-r(T_{max}, C_0)] = \frac{2U}{(-\Delta H)R}(T_{max} - T_M) \qquad (6\text{-}59)$$

The analysis is demonstrated in Figure 6.7. Two rate curves are shown,

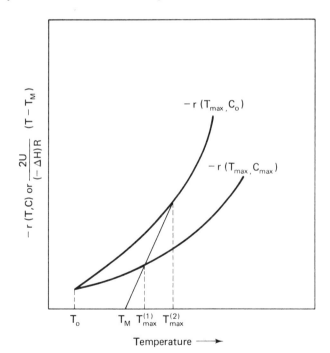

Figure 6.7 Parametric sensitivity analysis for a nonisothermal PFR (after P. Harriot, *Chem. Eng.*, **68**(10), 165 (1961); with permission of McGraw-Hill, Inc., New York)

$-r(T_{\max}, C_0)$, which we know, and $-r(T_{\max}, C_{\max})$, which we do not know but can be assured that for any temperature will lie below the value for $-r(T_{\max}, C_0)$. Heat removal is represented by a straight line whose location and slope are determined by T_M and U. $T_{\max}(2)$ is the design specified maximum based on $-r(T_{\max}, C_0)$, while $T_{\max}(1)$ is that actually obtained. Since the actual rate is smaller than $-r(T_{\max}, C_0)$, the conservative result $T_{\max}(2) > T_{\max}(1)$ is obtained. Artful compromise is required in such procedures, since overly conservative policies on U or T_M can essentially result in quenching the reaction; on the other hand, the analysis is based on steady-state balances and does not anticipate dynamic effects produced by perturbation of operating conditions. It has been suggested that maximum security in design can be obtained by making the straight line for heat transfer tangent to $-r(T_{\max}, C_0)$ at the specified value for $T_{\max}$. This yields the following specification, assuming Arrhenius temperature dependence:

$$(T_{\max} - T_M) = \frac{RT_{\max}^2}{E} \tag{6-60}$$

Further discussion of these approaches to parametric sensitivity has been given by Harriot [P. Harriot, *Chem. Eng.*, **68**(10), 165 (1961)].

The procedure described above has been criticized because it involves an a priori specification of an allowable maximum temperature, a procedure that may be somewhat arbitrary in many cases. In the work of Van Welsenaere and Froment previously cited, an attempt was made to eliminate this arbitrariness from the analysis of parametric sensitivity and base the approach on the nature of concentration/temperature trajectories within the reactor. For first-order kinetics let us write the heat-balance equation at the locus of the hot spot:

$$u\frac{dT}{dz} = 0 = \frac{(-\Delta H)k_0 e^{-E/RT}}{\rho C_p}C - \frac{2U(T - T_M)}{R\rho C_p}$$

from which we may obtain

$$C_{\max} = \frac{T_{\max} - T_M}{[(-\Delta H)R/2U]k_0 e^{-E/RT_m}} \tag{6-61}$$

This equation relates the concentration and temperature at the hot spot, and for a given reaction and reactor is a function of the wall temperature. The general shape of such curves is illustrated in Figure 6.8; these have been termed *maxima curves*. The temperature corresponding to the maximum of a maxima curve, obtained from $(dC/dT_m) = 0$, is

$$(T_{\max})_{\max} = \frac{1}{2}\left[\frac{E}{R} - \sqrt{\frac{E}{R}\left(\frac{E}{R} - 4T\right)}\right] \tag{6-62}$$

The particular importance of $(T_{\max})_{\max}$ is that C–T trajectories which intersect the maximum of the maxima curve are associated with parametrically sensitive behavior. One criterion thus proposed by Van Welsenaere and Froment

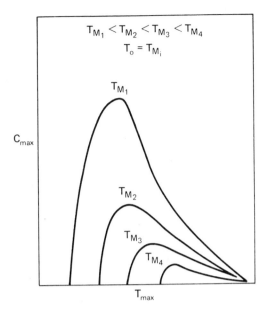

Figure 6.8 Shapes of a series of typical maxima curves (after R. J. Van Welsenaere and G. F. Froment, *Chem. Eng. Sci.*, **25**, 1503 (1970); with permission of Pergamon Press, Ltd., London)

for parametric sensitivity is that the "trajectory going through the maximum of the maxima curve is considered as critical and therefore as the locus of the critical inlet conditions for concentration and temperature corresponding to a given wall temperature...." Now in many cases $T_0 = T_M$, so the specification of a "critical" inlet condition means that for inlet concentration. A lower limit (conservative) on this may be established on the basis of an adiabatic trajectory from the maximum curve, since this will automatically ensure an inlet concentration less than that which could be tolerated when heat transfer is allowed. Thus,

$$[C_0(\text{critical}) - (C_{\max})_{\max}](-\Delta H) = \rho C_p[(T_{\max})_{\max} - T_0]$$

or

$$C_0(\text{critical}) = (C_{\max})_{\max} + \frac{\rho C_p}{(-\Delta H)}[(T_{\max})_{\max} - T_0] \qquad (6\text{-}63)$$

where $(C_{\max})_{\max}$ is determined from equation (6-61) with $T_{\max} = (T_{\max})_{\max}$, and $(T_{\max})_{\max}$ is obtained from equation (6-62).

As pointed out at the beginning of this section, these analyses are limited to reactor models for which T_M is constant and independent of position. In other situations the reactor may be cooled by a countercurrent flow of coolant, and we would have a reactor/heat exchanger, as shown in Figure 6.9. The problem has been analyzed by Grens and McKean [E. A. Grens and

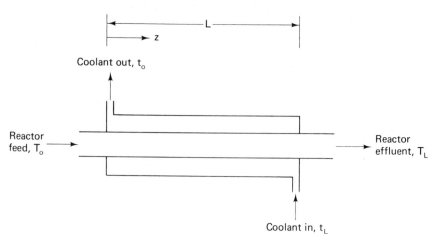

Figure 6.9 Countercurrent reactor/heat exhchanger

R. A. McKean, *Chem. Eng. Sci.*, **18**, 291 (1963)], who obtained an analytical solution for the temperature profiles of both reaction and coolant fluids for the case where reaction rate was independent of concentration and temperature. Although this may not be a very realistic assumption in many instances, the results do provide a "base case" for evaluating more complicated situations.

Two heat balances are required, one for reactant and one for coolant. For the reactant a slight modification of equation (4-121) will suffice:

$$\rho u C_p \frac{dT}{dz} = (-r)(-\Delta H) - \frac{2U}{R}(T - t) \tag{6-64}$$

while for the coolant

$$\rho_c u_c C_{p_c} \frac{dt}{dz} + \frac{2U}{R}(T - t) = 0 \tag{6-65}$$

with the boundary conditions $T = T_0$ at $z = 0$ and $t = t_L$ at $z = L$, as shown in Figure 6.9. A convenient set of nondimensional variables is

$$\zeta = \frac{t}{L} \qquad \phi = \frac{T - T_0}{\Delta T_0} \qquad \chi = \frac{t - T_0}{\Delta T_0}$$

where

$$\Delta T_0 = \frac{(-r)(-\Delta H)L}{\rho u C_p}$$

The two heat-balance equations then become

$$\frac{d\phi}{d\zeta} + \alpha(\phi - \chi) = 1 \tag{6-64a}$$

$$\frac{d\chi}{d\zeta} + \alpha\gamma(\phi - \chi) = 0 \tag{6-65a}$$

Two-Dimensional

Mixing-Cell Model

for Nonisothermal Reactors[1]

a. The mixing model

To develop the model we will begin with material and energy balances for a packed bed with individual particle diameter, dp, and no reaction (i.e., the mixing model). The geometry of such a bed in terms of the mixing-cell array is given in Figure 6A.1. Each cell, perfectly mixed, in the array is fed from two cells and feeds two cells in turn, thus allowing for radial as well as longitudinal mixing. We can define a characteristic cell dimension in terms of dp, in which case a holding time $\bar{t}$ is defined as

$$\bar{t} = \frac{dp}{v}$$

where v is the interstitial fluid velocity.[2] At a level i in the bed, then, the material balance is

$$C_{i-1} - C_i = \frac{dC_i}{dt}\left(\frac{dp}{v}\right) \tag{1}$$

where, in general,

$$C_i(0) = \bar{f}(i) \qquad 1 \leq i \leq N \quad \text{(function of position)}$$
$$C_0(t) = \bar{g}(t) \qquad t > 0 \qquad \text{(function of time)}$$

At each level i the horizontal stage is bounded by a plane at length $i\,dp$ and

---

[1] After H. A. Deans and L. Lapidus, *Amer. Inst. Chem. Eng. J.*, **6**, 656 (1960).
[2] Note this differs from nomenclature in the main text, where v is volumetric flow.

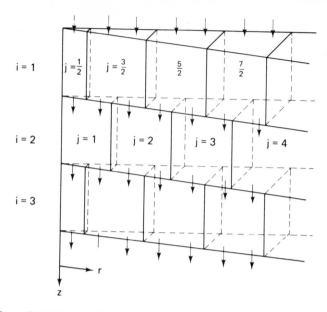

Figure 6A.1 Mixing-cell sequence in the model of Deans and Lapidus (after H. A. Deans and L. Lapidus, *Amer. Inst. Chem. Eng.*, **6**, 656 (1960); with permission of the American Institute of Chemical Engineers)

$(i - 1) dp$ from the entrance to the bed. Radial stages (Figure 6A.1) are bounded at a j by concentric cylindrical surfaces, radius $K(j - 1) dp$ and $Kj\, dp$. The index i is an integer, but j assumes noninteger values (multiples of $\frac{1}{2}$) for odd values of i. The radius in terms of particle diameters, M, is also taken to be an integer:

$$M = \frac{D_B}{(2dp)K} \tag{2}$$

where D_B is bed diameter and K is a scaling factor on the radial dimension such that $K\, dp$ is the mixing length in the radial dimension. [The K factor is introduced to bring the model into agreement with experimental values for the radial Peclet number as shown in Figures 5.10a and b. Deans and Lapidus report a correlation for K as

$$K = \left(\frac{8.2}{N_{\mathrm{Pe}_r}}\right)^{1/2}$$

For well-developed turbulence, $N_{\mathrm{Pe}_r} \approx 10$, then $K = 0.905$.] The volume per cell is

$$\bar{V}_{i,j} = \epsilon A_j\, dp \tag{3}$$

where ϵ is the bed porosity and A_j the cell interface area. In terms of the dimensions given,

$$A_j = \pi[K(j)\,dp]^2 - \pi[K(j-1)\,dp]^2$$
$$A_j = \pi K^2\,dp^2(2j-1) \tag{4}$$

Substituting this into equation (3),

$$\bar{V}_{i,j} = \bar{V}_j = \epsilon\,dp^3\pi K^2(2j-1) \tag{5}$$

Now at the centerline when $j = \frac{1}{2}, j - 1 = 0$, so

$$\bar{V}_{1/2} = \epsilon\pi K^2\,dp^3(\tfrac{1}{2})^2 \tag{6}$$

The total volumetric flow at (i,j) is

$$Q_{ij} = A_j\epsilon v \tag{7}$$

Now let us consider in detail an individual mixing cell as depicted in Figure 6A.2. Concentration in the cell is $C_{i,j}$ and, in accordance with the perfect mixing assumption, is the concentration in streams 3 and 4 leaving the cell. The problem is to determine the concentrations of streams 1 and 2 entering the cell. Let the total throughput in this stage be Q_j; then

$$\text{Stream 1:}\quad C_{i-1,j-1/2}$$
$$\text{Stream 2:}\quad C_{i-1,j+1/2}$$

The average concentration is

$$\phi_{i-1,j} = \frac{Q_1 C_{i-1,j-1/2} + Q_2 C_{i-1,j+1/2}}{Q_j} \tag{8}$$

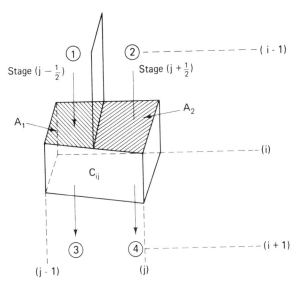

Figure 6A.2 Individual mixing cell in the matrix

where

$$Q_j = Q_1 + Q_2 = \epsilon \pi K^2 \, dp^2 v[j^2 - (j-1)^2] \tag{9}$$

Now we form the ratio (Q_1/Q_2):

$$\frac{Q_1}{Q_2} = \frac{A_1}{A_2} = \frac{(j-\frac{1}{2})^2 - (j-1)^2}{j^2 - (j-\frac{1}{2})^2} = \frac{j - \frac{3}{4}}{j - \frac{1}{4}} \tag{10}$$

Solving equations (9) and (10) for Q_1 and Q_2 and resubstituting in equation (8) gives

$$\phi_{i-1,j} = \frac{(j - \frac{3}{4})C_{i-1,j-1/2} + (j - \frac{1}{4})C_{i-1,j+1/2}}{2j - 1} \tag{11}$$

For the case where $j = \frac{1}{2}$ we are again involved with symmetry about the centerline. Here we say that $(j - \frac{3}{4}) = 0$, and the value of $\phi_{i-1,1/2}$ is

$$\phi_{i-1,1/2} = C_{i-1,1} \tag{12}$$

At the wall,

$$\phi_{i-1,M} = C_{i-1,M-1/2} \tag{13}$$

which is required by the continuum argument that $(dC/dr) = 0$ at the wall.

The heat balance may be written in a similar manner, for which in our pseudo-homogeneous model the solid and fluid temperatures are taken to be the same. Thus, corresponding to $\phi_{i-1,j}$ we may define an average temperature $\psi_{i-1,j}$ as

$$\psi_{i-1,j} = \frac{(j - \frac{3}{4})T_{i-1,i-1/2} + (j - \frac{1}{4})T_{i-1,j+1/2}}{2j - 1} \tag{14}$$

The entire mixing-cell model can then be represented in terms of two very simple equations:

$$\phi_{i-1,j} - C_{i,j} = \frac{dC_{i,j}}{dt}\left(\frac{dp}{v}\right) \tag{15}$$

$$\psi_{i-1,j} - T_{i,j} = \beta \frac{dT_{i,j}}{dt}\left(\frac{dp}{v}\right) \tag{16}$$

where

$$\beta = 1 + \frac{C_s \rho_s (1 - \epsilon)}{C_p \rho_f \epsilon}$$

C_s = heat capacity of the solid phase (mass)

ρ_s = density of the solid phase (mass)

C_p = heat capacity of the gas phase (molar)

ρ_f = density of the gas phase (molar)

$C_{i,j}$ and $T_{i,j}$ are dimensionless variables, with reference to inlet conditions. The form of equation (16), unlike the mass balance, must be modified at the wall to allow for heat transfer. At a cell next to the wall we can write, from equation (16) but adding a term for the heat transfer,

$$\psi_{i-1,M} - T_{i,M} + N_{ST_w}(\bar{T}_{w_i} - T_{i,M}) = \beta \frac{dT_{i,M}}{dt} \qquad (17)$$

where

$$N_{ST_w} = \frac{h_w A_{w_i}}{Q_M C_p \rho_f} = \text{Stanton number at the wall}$$

$$\bar{T}_{w_i} = \text{wall temperature}$$

$$Q_M = \text{total volumetric throughput}$$

$$A_{w_i} = 2\pi K M \, dp^2 \epsilon$$

The terms in A_{w_i} are explained as dp = length of cell, $2\pi KM \, dp$ = radius to outer wall, and ϵ = void volume, since heat transfer occurs only through the void volume. The total volumetric flow, Q_M, involved in equation (17) is given by

$$Q_M = A_M \epsilon v = \pi K^2 \, dp^2 (2M - 1)\epsilon v \qquad (18)$$

Thus,

$$N_{ST_w} = \frac{h_w A_{w_i}}{Q_M C_p \rho_f} = \frac{h_w (2\pi K M \, dp^2 \epsilon)}{\pi K^2 \, dp^2 (2M - 1)\epsilon v C_p \rho_f}$$

$$= \left(\frac{h_w}{G_M C_p}\right)\left(\frac{1}{K}\right)\left(\frac{2M}{2M - 1}\right) \qquad (19)$$

where G_M is the mass flow rate, $Q_M \rho_f$, and h_w is the heat-transfer coefficient. Now, equations (17) to (19) apply to the case where the cell wall corresponds to the vessel wall, but in alternate rows the outer cell will be fractional (odd values of i). These arrays may be treated in a similar manner:

$$A_{M_f} = \pi K^2 \, dp^2 [M^2 - (M - \tfrac{1}{2})^2]$$
$$= \pi K^2 \, dp^2 (M - \tfrac{1}{4}) \qquad (20)$$

and

$$Q_{M_f} = A_{M_f} \epsilon v = \pi K^2 \, dp^2 (M - \tfrac{1}{4}) v \epsilon \qquad (21)$$

so that

$$N_{ST_{w_f}} = \left(\frac{h_w}{G_M C_p}\right)\left(\frac{1}{K}\right)\left(\frac{2M}{M - \tfrac{1}{4}}\right) \approx 2N_{ST_w} \qquad \text{if } M \gg 1 \qquad (22)$$

We may now rewrite the energy balance at the wall, equation (17), so that it is of the same form as the general balance, equation (16):

$$\psi'_{i-1,M} - T_{i,M} = \beta' \frac{dT_{i,M}}{dt} \qquad (23)$$

where

$$\psi'_{i-1,M} = \frac{\psi_{i-1,M} + N_{ST_w} \bar{T}_{w_i}}{1 + N_{ST_w}} \qquad (24)$$

$$\beta' = \frac{\beta}{1 + N_{ST_w}}$$

Equations (15), (16), and (23) constitute the mixing-cell model for the un-steady-state, nonreactive bed. The problem again becomes an initial-value one, with conditions specified:

$$\left. \begin{array}{l} C_{i,j}(0) \\ T_{i,j}(0) \end{array} \right\} \quad \begin{array}{l} \text{state of } C \text{ and } T \text{ within the} \\ \text{bed at } t = 0 \end{array}$$

A solution procedure is simply developed as follows:

1. Inlet conditions $C_{0,j}(t)$ and $T_{0,j}(t)$ determine the values of $\phi_{0,j}$ and $\psi_{0,j}$ as a function of time.
2. $\phi_{0,j}(t)$ and $\psi_{0,j}(t)$ together with initial conditions $C_{i,j}(0)$ and $T_{i,j}(0)$ [or here $C_{1,j}(0)$ and $T_{1,j}(0)$, since we refer to the first row] allow solu-tion of equations (15) and (16) for $C_{1,j}$ and $T_{1,j}$ for all $t > 0$ (j is frac-tional, since i is odd on the first row).
3. The first-row solutions $C_{1,j}(t)$ and $T_{1,j}(t)$ determine $\phi_{1,j}(t)$ and $\psi_{1,j}(t)$, which in turn allow computation of $C_{2,j}(t)$ and $T_{2,j}(t)$ according to the procedures of step 2.
4. The computation is repeated for successive values of the index i to the bed exit.

An example of this type of calculation for a one-dimensional, isothermal bed is given at the end of this Appendix.

Note that the steady-state model is included in the description of equa-tions (15), (16), and (23), in which case the time derivatives disappear.

b. The reactor model

If it is assumed that variations in density and other physical parameters are small, equations (15) and (16) are readily modified to include reaction-rate terms. Thus, for component k of the reaction mixture in the (i, j)th stage:

$$\phi_k - C_k - [-r_k(C_k, \ldots, T)] = \frac{dC_k}{dt} \tag{25}$$

$$\psi - T - [-r_T(C_k, \ldots, T)] = \beta \frac{dT}{dt} \tag{26}$$

(where subscripts i and j have been omitted for clarity).

Let us simplify the model of equations (25) and (26) to the steady state and consider their detailed form for first-order kinetics. Then

$$(-r_k) = -kCe^{-E/T}$$

$$(-r_T) = k\lambda Ce^{-E/T}$$

where

$$\lambda = \frac{-(\Delta H_r)C_0}{C_p \rho_f T_0}$$

$$E = \frac{E_a}{RT_0}$$

$$k = k_0 \left(\frac{dp}{v}\right)$$

E_a = activation energy

k_0 = preexponential factor

the balance equations then become

$$\phi_{i-1,j} - C_{i,j}(1 + ke^{-E/T_{i,j}}) = \frac{dC_{i,j}}{dt} \tag{27}$$

$$\psi_{i-1,j} - T_{i,j} + \lambda C_{i,j} ke^{-E/T_{i,j}} = \beta \frac{dT_{i,j}}{dt} \tag{28}$$

for $1 \leq i \leq N$ and $0 \leq j \leq M$. Again, at the point where $j = M$, the effect of heat transfer at the wall must be included; using the same procedures employed previously,

$$\psi'_{i-1,M} - T_{i,M} + \lambda' C_{i,M} ke^{-E/T_{i,M}} = \beta' \frac{dT_{i,M}}{dt} \tag{29}$$

where

$$\psi'_{i-1,M} = \frac{\psi_{i-1,M} + N_{ST_w}\bar{T}_{w_i}}{1 + N_{ST_w}}$$

$$\lambda' = \frac{\lambda}{1 + N_{ST_w}}$$

$$\beta' = \frac{\beta}{1 + N_{ST_w}}$$

The same rule applies here as previously defined regarding the value to be used for N_{ST_w}, depending on whether i is even or odd.

Typical inlet and initial conditions are:

$$\phi_{0,j} = \phi_0(t) \qquad \text{inlet concentrations}$$
$$\psi_{0,j} = \psi_0(t) \qquad \text{inlet temperatures}$$
$$\bar{T}_{w_i} = \bar{T}_{w_i}(t) \qquad \text{wall temperatures}$$

and

$$C_{i,j}(0) = C(i,j) \qquad \text{initial concentrations}$$
$$T_{i,j}(0) = T(i,j) \qquad \text{initial temperatures}$$

In the steady state, we may write equations (27) and (28) as

$$\phi_{i-1,j} - C_{i,j}(1 + ke^{-E/T_{i,j}}) = 0 \tag{27a}$$

$$\psi_{i-1,j} - T_{i,j} + C_{i,j} \lambda e^{-E/T_{i,j}} = 0 \tag{28a}$$

Eliminating $C_{i,j}$ between this pair gives

$$\psi_{i-1,j} - T_{i,j} + \frac{\lambda \phi_{i-1,j}}{\dfrac{e^{-E/T_{i,j}}}{k} + 1} = 0 \tag{30}$$

This is essentially the same form of expression as equation (6-7) derived in the text for an adiabatic example. Recalling that $\psi_{i-1,j}$ and $\phi_{i-1,j}$ are numbers known from previous row solutions, we can then obtain $T_{i,j}$ from equation (30). Once this is obtained, $C_{i,j}$ is computed directly from equation (27a). For values of $j = M$ the same procedure is followed employing equations (27a) and (29).

Deans and Lapidus give also details on applications of the reactor model to unsteady-state simulation, and various complications arising from interphase transport limitations which extend beyond the scope of our present interests.

Example 1:   One-dimensional, Isothermal Mixing Model.

In the following we establish the equations for concentration response to a step-function concentration change, uniform across the radius, at time zero.
1. For $\phi_{0,j}$:

$$\phi_{0,j} = \frac{(j - \tfrac{3}{4})C_{0,j-1/2} + (j - \tfrac{1}{4})C_{0,j+1/2}}{2j - 1}$$

but $C_{0,j-1/2} = C_{0,j+1/2}$; therefore, $\phi_{0,j} = C_0$.
2. For the first row:

$$C_{0,j} - C_{1,j} = \frac{dC_{i,j}}{dt}$$

or

$$\frac{dC_{i,j}}{dt} + C_{i,j} = C_0 = \phi_0 \qquad \text{with } C_j = 0 \quad \text{at } t = 0$$

Thus, $C_{i,j} = C_0(1 - e^{-t})$.
3. For the second row:

$$\phi_{1,j} = \frac{(j - \tfrac{3}{4})C_{1,j-1/2} + (j - \tfrac{1}{4})C_{1,j+1/2}}{2j - 1}$$

Taking some values of j (j integer for i even),
$j = 1$:

$$\phi_{1,1} = \frac{(\tfrac{1}{4})C_0(1 - e^{-t}) + (\tfrac{3}{4})C_0(1 - e^{-t})}{1} = C_0(1 - e^{-t})$$

$j = 2$:

$$\phi_{1,2} = \frac{(\tfrac{5}{4})C_0(1 - e^{-t}) + (\tfrac{7}{4})C_0(1 - e^{-t})}{3} = C_0(1 - e^{-t})$$

$j = 3$:

$$\phi_{1,3} = \frac{(\tfrac{9}{4})C_0(1 - e^{-t}) + (\tfrac{11}{4})C_0(1 - e^{-t})}{5} = C_0(1 - e^{-t})$$

and so on. Thus,

$$\phi_{1,j} - C_{2,j} = \frac{dC_{2,j}}{dt}$$

or

$$\frac{dC_{2,j}}{dt} + C_{2,j} = C_0(1 - e^{-t})$$

4. Similar procedures yield differential equations giving higher-order responses for each stage i. The model here is formally analogous to the CSTR-sequence mixing model.

Example 2:   Sample Calculation Results for Two-dimensional Nonisothermal Reactor Model.

Deans and Lapidus present example calculations for the nonisothermal steady state using the following values of parameters:

$$2M = 11 \qquad N = 60 \qquad \lambda = 1 \qquad \frac{e^E}{k} = 250$$

$$\beta = 2.5 \qquad N_{ST_w} = 2 \qquad E = 10$$

The results of the computation are shown in Figure 6A.3 for longitudinal temperature and concentration profiles at various radial positions. In addition to demonstrating the reasonableness of the model in terms of the typical shapes of the profiles generated, the figures show also that significant radial as well as longitudinal gradients may be encountered in nonisothermal operation. It may be, then, often important to incorporate the two-dimensional approach in our nonideal reactor models, or at least to determine what level of approximation is involved in using a one-dimensional approach, as has been discussed in the text.

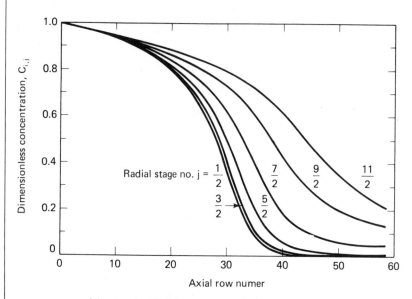

(a) Longitudinal Concentration Profiles at Various Radial
Positions, Static System

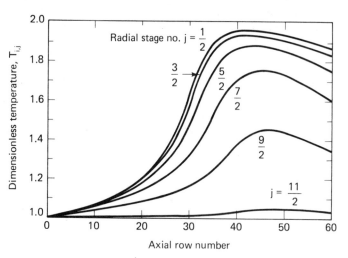

(b) Longitudinal Temperature Profiles at Various Radial
Positions, Static System

Figure 6A.3 Results of a typical steady-state simulation with the mixing-cell model of Deans and Lapidus (after H. A. Deans and L. Lapidus, *Amer. Inst. Chem. Eng. J.*, **6**, 663 (1960); with permission of the American Institute of Chemical Engineers)

Reactions in 7
Heterogeneous Systems

With the exception of the discussion of reactions on surfaces in chapter 3 and occasional references to fixed-bed reactors in chapter 5, we have been concerned exclusively with reactions in homogeneous phases and reactor models corresponding. Although these models are not completely restricted in application to homogeneous reactions, as we have discussed in chapter 6, there are properties of reactions in heterogeneous systems which are sufficiently important to warrant separate treatment.

Ordinarily two phases are involved: gas/solid, gas/liquid, or liquid/solid. Recent developments in reactor technology have also included three-phase systems, gas/liquid/solid, but these are somewhat beyond the scope of our present interests. The most important case of heterogeneous reaction must be the gas/solid system, which is typical of most catalytic processes; fortunately, the same general principles pertain to the analysis of all two-phase systems, so separate developments on a case-by-case basis are not required.

The primary feature of heterogeneous reactions is that the purely physical problem of transporting reactants and products between phases is appended to the chemical transformation. Stated alternatively, a physical rate process (transport) occurs in series with a chemical rate process (reaction). The overall behavior of the reaction system depends on the relative magnitude of these rate processes, not only in terms of the net kinetics of transformation but also with regard to the selectivity in complex reactions.

In this chapter we shall develop the theory on a basic level for reactions in some two-phase systems, then discuss applications to reactor modeling and design for some specific examples.

7.1 Reactions in Gas/Solid Systems[1]

Heterogeneous catalytic reactions by their nature involve the serial transport/ reaction rate process, since the reaction materials must be transported and removed from the surface site where the chemical transformation occurs. Much of the basic theory rests on the simplified picture of the combined transport and reaction process shown in Figure 7.1. The catalyst particle, containing a large specific surface area per unit volume, is bathed in reaction mixture as illustrated for a reactant A molecule and a product B molecule.

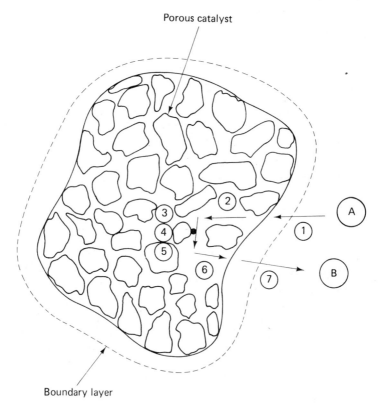

Figure 7.1 Sequential steps involved in a catalytic reaction

[1]The development here is generally applicable for liquid/solid catalytic reactions as well, although the magnitude of various parameters may be considerably different.

For the overall transformation, the following sequence is necessary:

1. Diffusion of A through a boundary layer or "film" adjacent to the external surface of the catalyst.
2. Diffusion through the porous interior of the catalyst to the point at which adsorption/reaction occurs.

3, 4, 5. Sequence of adsorption, surface reaction, and desorption, the details of which determine the intrinsic kinetics of the catalytic reaction.

6. Diffusion of B through the porous structure to the external surface.
7. Diffusion of B through the external boundary layer into the bulk gas phase.

Now if the intrinsic rate of the surface reaction steps 3 to 5 is slow, the rate of transport through the surrounding media would easily be great enough to supply instantaneously a molecule of A as needed for reaction and to remove the product B molecule when it is formed. Thus, when the intrinsic kinetics are slow, the overall transport/reaction process occurs without the establishment of significant concentration gradients within the porous catalyst particle or in the surrounding boundary layer. When the intrinsic kinetics are of similar magnitude to the mass transport rates, gradients in concentration will be required to produce the required rate of supply and removal of reactants and products. It is convenient to visualize relative mass transfer and chemical reaction rates in terms of these gradients as, indeed, they have a profound effect on the behavior of the catalyst as a whole. In Figure 7.2 is given a simple representation of gradients for several cases of relative mass transfer/chemical reaction rates. Since these gradients are established when transport rates become finite, the net effect is to reduce the overall rate of reaction due to the lower incident concentration of reactant within the catalyst as compared to surface (or bulk) concentrations. The net activity of the catalyst is diminished, and it is common to define this quantitatively in terms of the catalytic *effectiveness factor*, given by

$$\eta = \frac{\text{observed rate of reaction}}{\text{rate of reaction at surface concentration}} \qquad (7\text{-}1)$$

The definition of equation (7-1) does not envision differences between bulk and surface concentrations, a point that will be discussed later. Here we shall first treat the problem of transport limitations within the porous matrix (intraphase), then the combination of boundary-layer (interphase) transport with the intraphase effects.

Except for the discussion of some specific forms of complex rate equations and references to some specific studies of various reactions, we must again be content in the following with those ubiquitous reactants A and B and occasionally C, for purposes of selectivity. The rationale for our treatment is

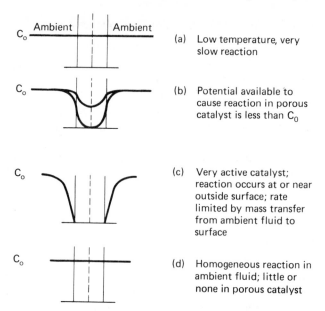

(a) Low temperature, very slow reaction

(b) Potential available to cause reaction in porous catalyst is less than C_0

(c) Very active catalyst; reaction occurs at or near outside surface; rate limited by mass transfer from ambient fluid to surface

(d) Homogeneous reaction in ambient fluid; little or none in porous catalyst

Figure 7.2 Concentration profiles in a porous catalyst under different reaction regimes (after C. N. Satterfield, *Mass Transfer in Heterogeneous Catalysis*, © 1970; with permission of MIT Press, Cambridge, Mass.)

that, while there are tangible chemical objectives ultimately involved, mass- and energy-transport processes are physical ones and illustration of their effects is best served by using the conventions of simple "nonchemical" kinetics.

Comprehensive treatments of the theory and applications of diffusion and chemical reaction have been given in the classical work of Frank-Kamenetskii (D. A. Frank-Kamenetskii, *Diffusion and Heat Transfer in Chemical Kinetics*, 2nd ed., Plenum Press, New York, 1969), by Satterfield (C. N. Satterfield, *Mass Transfer in Heterogeneous Catalysis*, MIT Press, Cambridge, Mass., 1970), and very recently by Aris (R. Aris, *The Mathematical Theory of Diffusion and Reaction in Permeable Catalysts*, Clarendon Press, Oxford, 1975).

a. Steady-state diffusion and chemical reaction in porous catalysts under isothermal conditions

First considerations in the modern literature of the problem of diffusion and reaction in porous catalysts were reported independently by Thiele [E. W. Thiele, *Ind. Eng. Chem.*, **31**, 916 (1939)], Damköhler [G. Damköhler, *Der Chemie-Ingenieur*, **3**, 430 (1937)], and Zeldovich [Ya. B. Zeldovich, *Acta Phys.-Chim. USSR*, **10**, 583 (1939)], although the first solution to the mathematical problem was given by Jüttner in 1909 [F. Jüttner, *Z. Phys. Chem.*, **65**, 595 (1909)]. Consider the porous catalyst in the form of a flat slab of semi-infinite dimension on the surface, and of half-thickness W as shown in Figure 7.3. A first-order, irreversible chemical reaction A $\longrightarrow$ B is catalyzed within the porous matrix with an intrinsic rate $(-r)$. We assume that the mass-transport process in one direction through the porous structure may be represented by a diffusion-type expression, there is no net convective contribution, and that the medium is isotropic. For this case, a steady-state mass balance over the differential volume element dz in Figure 7.3, carried out in the same manner as discussed previously, yields

$$D_{\text{eff}} \frac{d^2 C_A}{dz^2} - (-r) = 0 \qquad (7\text{-}2)$$

This model represents the catalyst as a pseudo-homogeneous medium in which the effective transport coefficient for the reactant, D_{eff}, is in general a complex function of the porous structure and composition, a topic we shall treat subsequently. It is normally a sufficient approximation to treat D_{eff} as a constant. The boundary conditions corresponding to equation (7-2) specify conditions at the surface and a symmetry at the center:

$$\begin{aligned} z &= W & C_A &= C_{A_0} \\ z &= 0 & \frac{dC_A}{dz} &= 0 \end{aligned} \qquad (7\text{-}3)$$

For first-order kinetics, $-r = kC_A$ and equation (7-2) may be written in non-dimensional form as

$$\frac{d^2 f}{d\zeta^2} - W^2 \left(\frac{k}{D_{\text{eff}}} \right) f = 0 \qquad (7\text{-}4)$$

where

$$f = \frac{C_A}{C_{A_0}}$$

$$\zeta = \frac{z}{W}$$

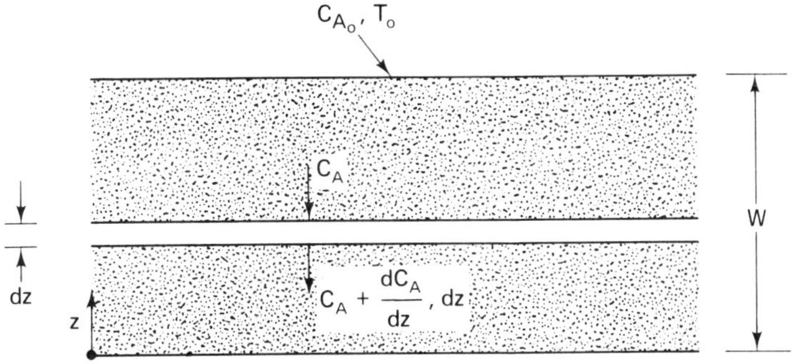

Figure 7.3 Diffusion and reaction in a semi-infinite slab of porous catalyst

The quantity $W^2(k/D_{eff})$ relates the relative magnitudes of the rate parameters for chemical reaction and mass transport and hence is a characteristic property of the simultaneous transport/reaction problem. It is most often termed the Thiele modulus, ϕ, defined by

$$\phi^2 = W^2(k/D_{eff}) \tag{7-5}$$

The solution of equation (7-4) gives for the concentration profile of reactant A through the slab:

$$f = \frac{C_A}{C_{A_0}} = \frac{\cosh(\phi\zeta)}{\cosh\phi} \tag{7-6}$$

The total rate of reaction can be obtained via Fick's law for the mass flux at the external surface using the concentration gradient evaluated at that point from equation (7-6). For unit surface area,

$$\text{total rate} = N_A = -D_{eff}\left(\frac{dC_A}{dz}\right)_W$$

This result is

$$N_A = \frac{D_{eff}C_{A_0}}{W}\phi \tanh\phi \tag{7-7}$$

We recall that the net effect on the intrinsic activity of the catalyst caused by the presence of a concentration gradient is defined in terms of the effectiveness factor, equation (7-1). The overall rate in the absence of gradients is $kC_{A_0}W$ (for unit surface area), so

$$\eta = \frac{D_{eff}C_{A_0}\phi \tanh\phi}{W \cdot kC_{A_0}W} = \frac{\tanh\phi}{\phi} \tag{7-8}$$

The net rate of reaction in the catalyst is then given by

$$(-r)_{net} = kC_{A_0}\eta = \frac{\tanh\phi}{\phi}kC_{A_0} \tag{7-9}$$

When the diffusional influence is very pronounced, that is, when $D_{eff} \ll k$, ϕ becomes large and $\tanh\phi$ approaches unity. The effectiveness factor for this *strong diffusion limit* is

$$\eta = \frac{1}{\phi}$$

The results above are specific to the geometry employed in Figure 7.3. For spherical coordinates the corresponding solution is

$$\eta = \frac{3}{\phi_S^2}(\phi_S \coth\phi_S - 1) \tag{7-10}$$

with

$$\phi_S = R_p\sqrt{k/D_{eff}}$$

and for cylindrical geometry (infinite in extent along the axis):

$$\eta = \frac{2I_1(\phi_c)}{\phi_c I_0(\phi_c)} \tag{7-11}$$

with ϕ_c defined similarly to ϕ_S, R_p being the radius in both cases. In equation

(7-11) I_0 and I_1 are modified Bessel functions of the first kind. The apparent geometric dependence of these results for η can be removed for most engineering applications by defining a characteristic dimension of the particular geometry involved as the ratio of volume $\bar{V}$ to external surface area S_A [R. Aris, *Chem. Eng. Sci.*, **6**, 292 (1957)]. For a sphere $(\bar{V}/S_A)$ is $(R_p/3)$, while for the slab $(\bar{V}/S_A)$ is W; hence, the values of ϕ employed in equations (7-8) and (7-10) differ by a factor of 3 if one wishes to compare corresponding conditions in the two geometries. Figure 7.4 gives some solutions for effectiveness factors plotted in accord with the $(\bar{V}/S_A)$ rule. The maximum difference in the two is about 10% in η for $2 \leq \phi_s \leq 6$. In subsequent discussion we shall present various results, some for spherical and some for slab geometries, without making special distinction between the two. As one would infer, cylindrical geometry gives results intermediate between these two limits.

When $(-r)$ is given as an integer power law, results completely analogous to (7-8) to (7-11) may be obtained with the Thiele modulus defined as

$$\phi = W\left(\frac{kC_{A_0}^{n-1}}{D_{\text{eff}}}\right)^{1/2} \tag{7-12}$$

where n is the order of reaction. The results of calculations using this definition are also depicted in Figure 7.4. One point of interest (and caution) is the behavior of a zero-order reaction; in this case there are two regions of solution, one in which C_A is finite at $z = 0$ and the second in which C_A becomes zero at some point $z < W$. In the former case $\eta = 1$, effectiveness is independent of ϕ, and although gradients exist, they do not affect the rate of reaction.

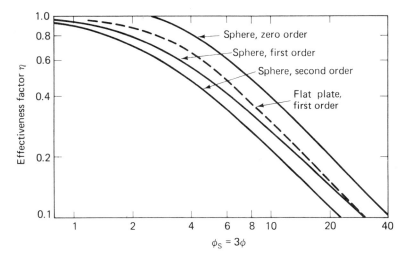

Figure 7.4 Effectiveness factors for power-law kinetics and various geometries: ϕ_S (sphere) $= 3\phi$ (slab) (after C. N. Satterfield, *Mass Transfer in Heterogeneous Catalysis*, © 1970; with permission of MIT Press, Cambridge, Mass.)

When C_A becomes zero at some interior point, effectiveness must decrease since a portion of the catalyst is not contacted with the reactant. [See V. W. Weekman, Jr. and R. L. Gorring, *J. Catalysis*, **4**, 260 (1965)]. The analysis of reaction systems in which the kinetics are not simple, Langmuir–Hinshelwood for example, extends beyond elementary methods and has been treated by Chu and Hougen [C. Chu and O. A. Hougen, *Chem. Eng. Sci.*, **17**, 176 (1962)] and Roberts and Satterfield [G. W. Roberts and C. N. Satterfield, *Ind. Eng. Chem. Fundls.*, **4**, 288 (1965); **5**, 317 (1966)].

b. Implications with respect to temperature dependence and reaction order

We have previously defined the apparent activation energy of a reaction as

$$\frac{E_a}{RT^2} = \frac{d \ln k}{dT} \tag{1-26}$$

or more generally as

$$E_a = R \frac{d \ln(-r)}{d(1/T)} \tag{7-13}$$

where $(-r)$ is the observed reaction rate. Now the effectiveness factor is written such that the observed rate is given as

$$(-r) = \eta \cdot f(T_0, C_0) \tag{7-14}$$

where $f(T_0, C_0)$ is the functional form of the rate expression under gradientless conditions. For a rate constant obeying the Arrhenius equation,

$$(-r) = \eta \cdot k^\circ \exp\left(\frac{-E}{RT}\right) \cdot f'(C_0) \tag{7-15}$$

substitution into equation (7-14) and performing the indicated differentiation gives

$$\frac{d \ln(-r)}{d(1/T)} = \frac{d \ln k}{d(1/T)} + \frac{d \ln \eta}{d(1/T)}$$

or

$$E_a = E + R \frac{d \ln \eta}{d \ln \phi} \cdot \frac{d \ln \phi}{d(1/T)} \tag{7-16}$$

For small values of ϕ, $(d \ln \eta/d \ln \phi) = 0$ and the apparent and Arrhenius activation energies are the same. For large values of ϕ this derivative approaches -1 and, assuming D_{eff} can also be approximately represented by an Arrhenius form,[1]

$$D_{eff} = D_{eff}^\circ \exp\left(\frac{-E_D}{RT}\right) \tag{7-17}$$

---

[1] This is not the theoretical temperature dependence of D_{eff}, as will be shown later; however, the approximation is useful within the present context. "Cudgel thy brains no more about it" (William Shakespeare, *Hamlet*).

then equation (7-16) gives

$$E_a = \frac{E + E_D}{2} \tag{7-18}$$

Normally, $E \gg E_D$ since diffusion is not very temperature-sensitive, so the apparent activation energy is approximately one-half the true value when pronounced concentration gradients are present.

In Figure 7.5 is plotted in Arrhenius form what might be observed for the reaction rate in a catalytic reaction as one continues to increase the temperature. At low temperatures the reaction is slow, mass transport is no limitation, and one observes the true activation energy. As the temperature increases, the rate of reaction increases more rapidly than the rate of mass transport, transport limitations appear, and the observed activation energies begin to fall off toward the ultimate limit given by equation (7-18). When both intra- and interphase gradients occur, the limits predicted by equations (7-16) and (7-18) are considerably modified. This will be treated later, as well as the details of the calculations by Weisz and Hicks given in Figure 7.5 for the combination of mass- and heat-transfer effects. Figure 7.6 gives some experimental results, obtained for the cracking of cumene on silica-alumina catalysts in which the diffusional modulus has been changed by variation of the catalyst pellet dimension [P. B. Weisz and C. D. Prater, *Advan. Catalysis*, **6**, 144 (1954)]. It is seen that the slope of the $\ell n\, r$ versus $1/T$ plot for the severely diffusionally limited pellets of 0.175-cm radius is approximately one-half that of the 0.0056-cm pellets.

A similar analysis can be conducted for the influence of mass transport on the apparent reaction order. For a system with intrinsic nth-order kinetics:

$$(-r) = \eta k C_0^n \tag{7-19}$$

the apparent order with respect to C_0 is

$$\frac{d \ell n\, (-r)}{d \ell n\, (C_0)} = n_a \tag{7-20}$$

Combination of equations (7-19) and (7-20) gives

$$n_a = n + \frac{d \ell n\, \eta}{d \ell n\, \phi} \cdot \frac{d \ell n\, \phi}{d \ell n\, C_0} \tag{7-21}$$

with ϕ as defined in equation (7-12) and D_{eff} independent of concentration. This gives

$$n_a = n + \frac{n - 1}{2} \cdot \frac{d \ell n\, \eta}{d \ell n\, \phi} \tag{7-22}$$

The apparent reaction order will lie somewhere between the true order and unity. For first-order reactions the diffusion process will have no effect on order, as is apparent from equation (7-22).

These modifications of kinetics, both with respect to activation energy and reaction order, when mass transport becomes rate-limiting have profound

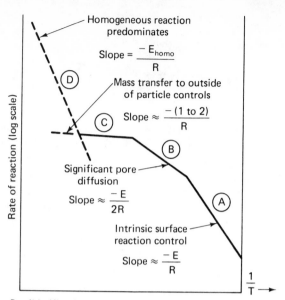

(a)   Possible Kinetic Regimes in a Gas-Phase Reaction Occurring on a Porous Solid Catalyst

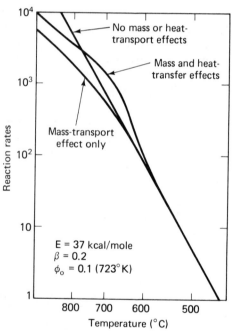

E = 37 kcal/mole
$\beta = 0.2$
$\phi_o = 0.1$ (723°K)

(b)   Modification of the Arrhenius Plot of Activity Due to Mass-Transport Effect, and Due to Both Mass and Heat-Transport Effect

Figure 7.5 Differing regimes of diffusion as they might affect an Arrhenius correlation (a) (after C. N. Satterfield, *Mass Transfer in Heterogeneous Catalysis*, © 1970; with permission of MIT Press, Cambridge, Mass.; (b) (after P. B. Weisz and J. S. Hicks, *Chem. Eng. Sci.*, **17**, 265 (1962); with permission of Pergamon Press, Ltd., London)

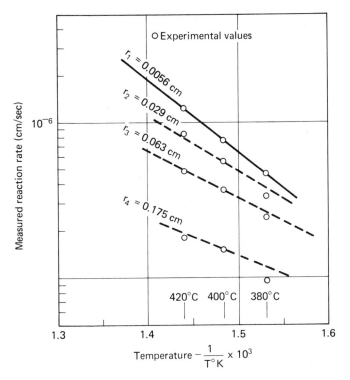

Figure 7.6 Variation in apparent activation energy with extent of diffusion for cumene cracking on SiO_2/Al_2O_3 (after P. B. Weisz and C. D. Prater, *Advan. Catalysis*, **6**, 144 (1954); with permission of Academic Press, Inc., New York)

implications in all aspects of catalysis and have quite justifiably been referred to as the "diffusional falsification" of kinetics. It is of importance for the experimentalist to be aware of such effects when investigating a reaction in the laboratory, and for the designer to account for them in formulating the kinetics pertinent to actual process conditions.

c. Nonisothermal theory

Since a large number of catalytic reactions involve thermal effects due to the heat of reaction (in many cases of technological interest, these are exothermic reactions) it may be in many cases not realistic to pursue expectations based on isothermal analysis. In the short development below, a direct proportion between temperature and concentration gradients is indicated [C. D. Prater, *Chem. Eng. Sci.*, **8**, 284 (1958)]; thus, in cases where significant concentration gradients exist and the reactions are not thermally neutral, it is reasonable to expect that thermal gradients will also exist. Under steady-state conditions we may write for the mass balance,

$$D_{eff}\frac{d^2C}{dz^2} - (-r) = 0 \tag{7-23}$$

and

$$k_{eff}\frac{d^2T}{dz^2} + (-\Delta H)(-r) = 0 \tag{7-24}$$

in which $(-\Delta H)$ is the heat of reaction per mole of reactant, and k_{eff} is an effective thermal conductivity in the porous matrix. Since $(-r)$ is the same in both expressions, we can write

$$\frac{(-\Delta H)D_{eff}}{k_{eff}}\frac{d^2C}{dz^2} = \frac{d^2T}{dz^2} \tag{7-25}$$

For surface conditions of C_0 and T_0, integration of equation (7-25) gives

$$T - T_0 = \frac{(-\Delta H)(D_{eff})}{k_{eff}}(C_0 - C) \tag{7-26}$$

which shows that the relationship between temperature and concentration is linear and is independent of geometry and reaction kinetics. The largest possible difference is when $C = 0$; then

$$(\Delta T)_{max} = \frac{(-\Delta H)D_{eff}}{k_{eff}}C_0 \tag{7-27}$$

It was long a point of some controversy as to what magnitudes of $(\Delta T)_{max}$ might be expected, with estimates ranging from hundredths to hundreds of degrees (on whatever scale); recent experiments indicate that internal ("intraphase") gradients may range up to the order of tens of degrees, but that the external thermal gradients are more important. [See, for example, C. McGreavy and D. L. Cresswell, *Chem. Eng. Sci.*, **24**, 608 (1969); J. P. G. Kehoe and J. B. Butt, *Amer. Inst. Chem. Eng. J.*, **18**, 347 (1972)]. We shall return to this point in due course.

Solution of the problem of intraphase nonisothermal effectiveness requires simultaneous solution of equations (7-23) and (7-24) using boundary conditions for temperature and concentration analogous to equation (7-3). Various methods, such as linear approximations [R. E. Schilson and N. R. Amundson, *Chem. Eng. Sci.*, **13**, 226, 237 (1961)] or perturbation analysis have been employed to obtain approximate analytical solutions, but a general treatment requires numerical solution. If equations (7-23) and (7-24) are cast into nondimensional form, two additional parameters appear:

$$\beta = \text{heat of reaction parameter} = \frac{(\Delta T)_{max}}{T_0} = \frac{(-\Delta H)D_{eff}}{k_{eff}T_0}C_0$$

$$\gamma = \text{activation energy parameter} = \frac{E}{RT_0}$$

As an example of typical results, Weisz and Hicks [P. B. Weisz and J. S. Hicks, *Chem. Eng. Sci.*, **17**, 265 (1962)] report the solution for the case of first-order reaction in spherical geometry. Some of these are shown in Figure 7.7 for a range of the parameters β, γ, and ϕ_S. The isothermal theory is

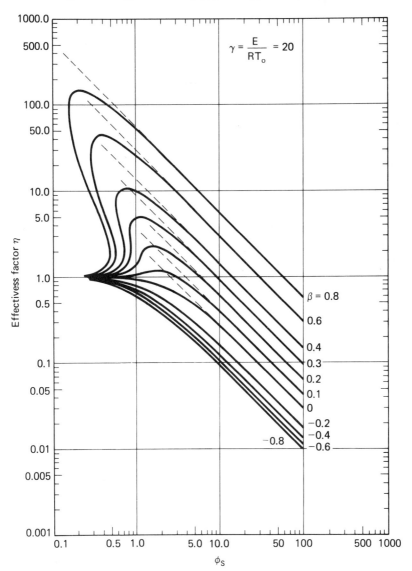

Figure 7.7 Nonisothermal effectiveness factor for first-order reaction in a sphere $\phi_S = R\sqrt{k/D_{eff}}$ (after P. B. Weisz and J. S. Hicks, *Chem. Eng. Sci.*, **17**, 265 (1962); with permission of Pergamon Press, Ltd., London) (Dotted lines are the asymptotic solution given by Petersen) (E. E. Petersen, *Chem. Eng. Sci.*, **17**, 987 (1962))

contained as a special case ($\beta = 0$) of the nonisothermal analysis. There are two important differences in the results shown in Figure 7.7 from the isothermal theory. First is the fact that the effectiveness factor can be greater than unity. This arises from the more pronounced dependence of reaction rate on temperature than on reactant concentration; hence, in certain cases it is possible for intraphase temperature gradients to "overcompensate" the concentration gradients with a corresponding increase in overall reaction rate. The second important difference, perhaps not so surprising in view of our previous experience with nonisothermal reaction systems, is the appearance of multiple steady states for certain ranges of β and ϕ_s in the figure. This can be explained physically on much the same basis as we have used in discussion of thermal effects on the CSTR, in terms of relative rates of heat generation and removal.

The qualitative picture of activity variation as a function of temperature is much the same in the nonisothermal case as we have discussed already for isothermal conditions. Again at sufficiently high temperature (large diffusional resistances) the apparent activation energy approaches a limit of half the true value. Since the curves of η versus ϕ_s are not monotonically decreasing for $\beta > 0$, however, the derivative in the relationship between observed and actual activation energies becomes greater than zero in certain temperature ranges and the apparent activation energy is greater than the actual value. An example of this is shown on the lower part of Figure 7.5 for the curve labeled "Mass- and heat-transfer effects."

A useful asymptotic method has been developed by Petersen [E. E. Petersen, *Chem. Eng. Sci.*, **17**, 987 (1962)] for the estimation of η for large values of ϕ_s. When $\phi_s \gg 1$ we may write the mass conservation equation in the form

$$\frac{d^2 f}{d\mu^2} + (-r') = 0 \qquad (7\text{-}28)$$

where

$$\mu = (1 - \zeta)\phi_s$$

$$\zeta = \frac{r}{R_p} \text{(spherical geometry)}$$

$$f = \frac{C}{C_0}$$

$$(-r') = \frac{-r(\zeta)}{-r(1)}$$

with the boundary conditions:

$$\mu = 0: \quad f = 1$$

$$\mu \longrightarrow \infty: \quad \frac{df}{d\mu} = 0 \qquad f = 0$$

The second condition here reflects that, under conditions of strong diffusional limitation, the reactant will be totally consumed in the outer layers of the particle. If we make the substitution $p = df/d\mu$, equation (7-28) may be directly integrated to

$$\frac{df}{d\mu} = -\sqrt{2}\left[\int_0^{f(\mu)} (-r')\,df\right]^{1/2} \qquad (7\text{-}29)$$

Normally one is interested in $(df/d\mu)$ at the surface, since this is required for the evaluation of surface flux and hence effectiveness. It can be shown that

$$\eta = \frac{1}{\phi_s^2}\left(\frac{df}{d\mu}\right)_{\mu=0}$$

so that

$$\eta = \frac{\sqrt{2}}{\phi_s}\left[\int_0^1 (-r')\,df\right]^{1/2} \qquad (7\text{-}30)$$

This approximation may be used for isothermal or nonisothermal systems. The dashed lines on Figure 7.7 are solutions obtained by this method; applications of the procedure are detailed by Petersen (E. E. Petersen, *Chemical Reaction Analysis*, p. 70, Prentice-Hall, Englewood Cliffs, N.J., 1965).

The curves presented in Figure 7.7 are functions of the two parameters β and γ, and as presented here a separate graph would be required for each value of γ of interest. It had been shown earlier [J. J. Carberry, *Amer. Inst. Chem. Eng. J.*, 7, 350 (1961)] that for first-order reactions and for the product $\beta\gamma$ from a range of -4 to 6, a single set of curves can be used to represent the nonisothermal effectiveness factor (second-order kinetics can also be managed in this way). Following this suggestion, Liu [S-L. Liu, *Amer. Inst. Chem. Eng. J.*, 16, 742 (1970)] has presented some empirical representations for η as follows (first-order irreversible reaction):

$$\text{For } \phi_s > 2.5: \quad \eta \approx \frac{\exp{(\delta/5)}}{\phi_s} \qquad (7\text{-}31a)$$

With the exception of the following parametric values:

$$\delta > 2.5: \quad \phi_s < 1.235 - 0.94\delta$$
$$\delta \le 2.5: \quad \phi_s < 1.820 - 0.32\delta$$

In this region

$$\eta = \exp{(0.14\phi_s\delta^{1.6})} - 1 + \frac{\tanh\phi_s}{\phi_s} \qquad (7\text{-}31b)$$

where $\delta = \beta\gamma$. For endothermic reactions,

$$\eta = \frac{\tanh\phi_s^*}{\phi_s^*} \qquad (7\text{-}31c)$$

where

$$\phi_s^* = \phi_s(1 - \delta)^{0.3}$$

d. Some intraphase selectivity problems

The effect of transport limitations on selectivity is in many instances far more pronounced than on activity. Important early work in this area for isothermal systems was conducted by Wheeler (A. Wheeler, *Catalysis*, vol. 2, P. H. Emmett, ed., Reinhold, New York, 1955) using the Types I, II, and III selectivity models we have employed in previous chapters here. To review, these are:

Type I:
$$A \longrightarrow B \quad (k_1)$$
$$L \longrightarrow M \quad (k_2)$$

Type II:
$$B \quad (k_1)$$
$$A$$
$$C \quad (k_2)$$

Type III:
$$A \longrightarrow B \longrightarrow C \quad (k_1, k_2)$$

The following analysis is presented for a given condition of reactant concentration (i.e., a point or differential selectivity), so that in reactor modeling the effects illustrated will be modified according to the conversion level.

Type I. In the absence of diffusion we may write

$$r_A = -k_1 C_{A_0}$$
$$r_L = -k_2 C_{L_0}$$

and a differential selectivity (slightly modified from equation (1-73) is

$$\frac{r_A}{r_A + r_L} = \frac{1}{1 + R/S_i} = \Delta \tag{7-32}$$

where S_i is the intrinsic selectivity, k_1/k_2, and $R = (C_{L_0}/C_{A_0})$. In the presence of an intraphase transport limit,

$$r_A = -\eta_A k_1 C_{A_0} \qquad r_L = -\eta_L k_2 C_{L_0}$$

and

$$\frac{r_A}{r_A + r_L} = \frac{1}{1 + \eta_L k_2 C_{L_0}/\eta_A k_1 C_{A_0}} = \Delta_{\text{diff}} \tag{7-33}$$

Under strong diffusion conditions for isothermal systems we have $\eta \rightarrow 1/\phi$, so equation (7-33) becomes

$$\Delta_{\text{diff}} = \frac{1}{1 + (k_2 C_{L_0}/\phi_L)(\phi_A/k_1 C_{A_0})} \tag{7-34}$$

and

$$\frac{\Delta_{\text{diff}}}{\Delta} = \frac{1 + (R/S_i)}{1 + R\phi_A/S_i\phi_L} \tag{7-35}$$

In this case $(\Delta_{\text{diff}}/\Delta)$ can be greater or less than unity, depending on the value of (ϕ_A/ϕ_L). For nearly equal effective diffusivities of the reactants $(\phi_A/\phi_L) \approx S_i^{1/2}$, equation (7-35) becomes

$$\frac{\Delta_{\text{diff}}}{\Delta} = \frac{1 + R/S_i}{1 + R/S_i^{1/2}} \tag{7-35a}$$

and the value of $(\Delta_{\text{diff}}/\Delta)$ depends on whether S_i is greater or less than unity. This is illustrated in Figure 7.8; for intrinsic selectivities greater than unity $(\Delta_{\text{diff}}/\Delta) < 1$ but not greatly so, particularly for small values of R; for S_i less than unity, the point selectivity is considerably increased in the presence of strong diffusion.

Type II. In this case, assuming equal-order kinetics exist for the two reaction paths, there can be no influence of diffusion on selectivity, since a common reactant is involved.

Type III. The first step in this sequence is given by the solution of equation (7-4). For reactant A

$$C_A = C_{A_0}\left[\frac{\cosh\left(\sqrt{k_1/D_{\text{eff,A}}}\ z\right)}{\cosh \phi}\right] \tag{7-36}$$

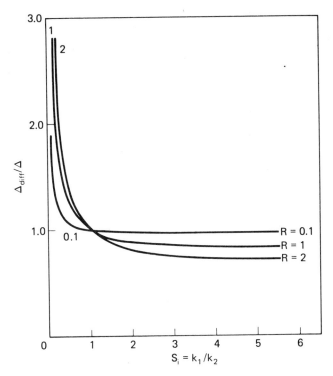

Figure 7.8 Influence of diffusion on selectivity in a Type I reaction system

and for B

$$D_{\text{eff,B}} \frac{d^2 C_B}{dz^2} - k_2 C_B = -k_1 C_A \tag{7-37}$$

where C_A in equation (7-37) is given by (7-36). The differential selectivity, defined before as the ratio of the rate of production of intermediate to the rate of consumption of reactant, is

$$\Delta_{\text{diff}} = \frac{r_B}{r_A} = \frac{D_{\text{eff,B}} (dC_B/dz)_W}{D_{\text{eff,A}} (dC_A/dz)_W} \tag{7-38}$$

Solving equation (7-37) for $C_B = f(z)$ and obtaining the derivatives involved in equation (7-38) gives, for conditions of large diffusional effects

$$\Delta_{\text{diff}} = \frac{R}{\sqrt{S_i}} - \frac{\sqrt{S_i}}{1 + \sqrt{S_i}} \tag{7-39}$$

where the ratio of effective diffusivities has been taken as unity and $R = C_{B_0}/C_{A_0}$. In the absence of diffusional limits

$$\Delta = -1 + \frac{R}{S_i} \tag{7-40}$$

so that

$$\frac{\Delta_{\text{diff}}}{\Delta} = \frac{\sqrt{S_i}[R(1 + \sqrt{S_i}) - S_i]}{(R - S_i)(1 + \sqrt{S_i})} \tag{7-41}$$

The differential selectivities vary with degree of conversion as represented by the value of R (c.f., Figure 7.8), so to obtain a complete picture we should also compare the behavior of diffusionally and non-diffusionally limited systems as reaction proceeds. The Type III selectivity problem is of particular interest in this regard, and has been treated in the original work of Wheeler. If we integrate equation (7-40) (no diffusional limitation) over a range of conversions

$$f_B = \frac{S_i}{S_i - 1}(1 - f_A)[(1 - f_A)^{(1-S_i)/S_i} - 1] \tag{7-42}$$

where f_B is the fraction of initial A reacted to B and f_A is the total fraction of A reacted. For the diffusion limited case, similar integration of equation (7-39) gives

$$f_B = \frac{S_i}{S_i - 1}(1 - f_A)[(1 - f_A)^{(1-\sqrt{S_i})/\sqrt{S_i}} - 1] \tag{7-43}$$

Plots of the comparative differential selectivities [equations (7-39) and (7-40)] and the overall selectivities [equations (7-42) and (7-43)] are given in Figure 7.9. In the region of primary interest, for $S_i > 1$, both Δ and Δ_{diff} assume negative values, indicative of a net production of the intermediate B (the negative sign arises from the difference in the directions of the fluxes of A and B at the catalyst surface). When $S_i > 1$, the effectiveness factor for A is less than that for B, assuming equal diffusivities, so that both point and

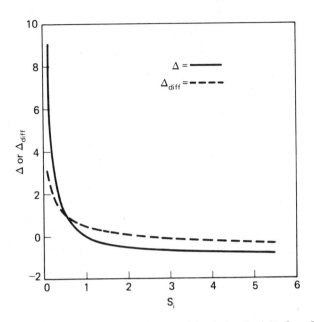

Figure 7.9a Variation of Δ and Δ diff with intrinsic selectivity for a Type III reaction, $R = 1$

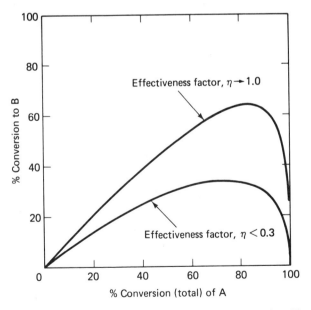

Figure 7.9b Variation of overall selectivity with conversion in a Type III reaction, $S_i = 4.0$ (from *Catalysis*, vol. 2, edited by P. H. Emmett, © 1955 by Litton Educational Publishing, Inc.; reprinted by permission of Van Nostrand Reinhold Company, New York)

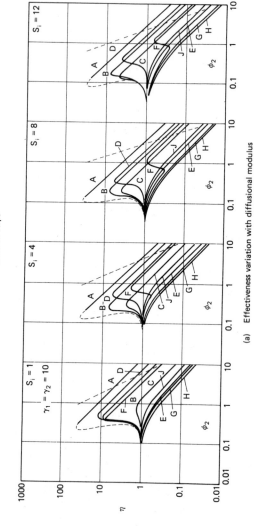

Reaction System	β_1	β_2
A	0.80	0.80
B	0.80	0.00
C	0.80	−0.40
D	0.00	0.80
E	0.00	−0.40
F	−0.40	0.80
G	−0.40	0.00
H	−0.40	−0.40
J	0.00	0.00

Thermal parameters of the two reaction steps

(a) Effectiveness variation with diffusional modulus

Figure 7.10 Heat and mass diffusion effects on selectivity in a Type III reaction (after J. B. Butt, *Chem. Eng. Sci.*, **21**, 275 (1966); with permission of Pergamon Press, Ltd., London)

overall selectivities for production of intermediate are adversely affected. In both illustrations of Figure 7.9 the loss in point selectivity or in maximum yield of intermediate is on the order of 50%, so these selectivity problems are not trivial.

For nonisothermal systems a point-selectivity analysis has been given for Type II reactions by Østergaard (K. Østergaard, *Proc. III Int. Congr. Catalysis*, North-Holland, Amsterdam, 1965) and for Type III reactions by Butt [J. B. Butt, *Chem. Eng. Sci.*, **21**, 275 (1966)]. In both instances the selectivity is computed from equation (7-38), where the concentration derivatives are evaluated by numerical solution of the corresponding conservation equations. For example, for the Type III reaction the equations to be solved are

$$D_{eff,A} \frac{d^2 C_A}{dz^2} - (-r_A) = 0$$

$$D_{eff,B} \frac{d^2 C_B}{dz^2} + (-r_A) - (-r_B) = 0 \qquad (7\text{-}44)$$

$$k_{eff} \frac{d^2 T}{dz^2} + (-\Delta H_A)(-r_A) + (-\Delta H_B)(-r_B) = 0$$

with corresponding parameters β_1, β_2, γ_1, and γ_2 pertaining to the individual reaction steps. Some results for the Type III reaction are illustrated in Figure 7.10. The catalytic effectiveness (a) behaves in much the same manner as for simple reactions, and the selectivity (b) is affected very strongly by both the heat- and mass-transport parameters. The shift from net negative to net positive values of Δ_{diff} illustrated in part (c) indicates that at higher ϕ the intermediate is reacting as fact as it is produced and whatever intrinsic selectivity for production of intermediate existed has been annihilated. We see also on Figure 7.10 that the locus of the region of multiplicity in a Type III reaction is strongly dependent on the relative β values for the two steps.

e. Diffusion and chemical reaction: the interphase/intraphase problem

To this point we have dealt only with transport effects within the porous catalyst matrix (intraphase), and the appropriate mathematics have been worked out for boundary conditions that specify concentration and temperature at the catalyst surface. In actual fact, external boundaries often exist which offer "resistance" to heat and mass transport, as shown in Figure 7.1, and the surface conditions of temperature and concentration may differ substantially from those measured in the bulk fluid. In fact, if internal gradients of temperature exist, interphase gradients in the boundary layer must also exist because of the relative values of the thermal conductivities [see, for example, J. J. Carberry, *Ind. Eng. Chem.*, **58**(10), 40 (1966)].

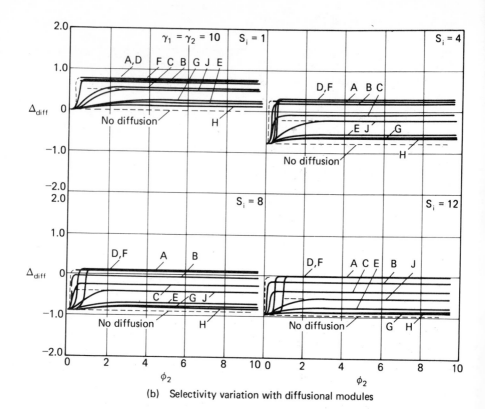

(b) Selectivity variation with diffusional modules

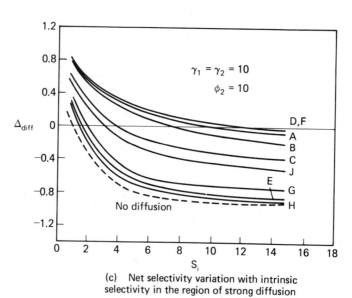

(c) Net selectivity variation with intrinsic selectivity in the region of strong diffusion

Figure 7.10 Heat and mass diffusion effects on selectivity in a Type III reaction

Reformulation of the problems previously examined to account for interphase gradients requires essentially only the change of the surface boundary conditions. Assuming that we may use simple mass- and heat-transfer coefficients as the rate coefficients characteristic of interphase transport, the boundary conditions for mass and energy conservation equations become

$$z = 0: \quad \frac{dC}{dz} = 0 \qquad z = W: \quad D_{eff}\left(\frac{dC}{dz}\right) = k_m(C_0 - C_s)$$

$$z = 0: \quad \frac{dT}{dz} = 0 \qquad z = W: \quad k_{eff}\left(\frac{dT}{dz}\right) = h(T_0 - T_s) \tag{7-45}$$

where C_0 and T_0 refer to bulk conditions and T_s and C_s to surface conditions.

In the isothermal case we may solve the problem without attempting solution of the mass balance with the new boundary conditions if the kinetics are simple. Equate the rate of mass transport at the external surface to the rate of reaction:

$$k_m S_A(C_0 - C_s) = (-r)\bar{V} \tag{7-46}$$

where $(-r)$ is the observed rate of reaction per unit volume of catalyst, $\bar{V}$ the catalyst volume, and S_A the external surface area. For first-order reactions,

$$k_m S_A(C_0 - C_s) = \eta k C_s \bar{V} \tag{7-47}$$

Solving equation (7-47) for the unknown surface concentration and resubstituting into the expression for the rate of reaction $(-r = \eta k C_s)$ yields

$$(-r) = \frac{\eta k C_0}{1 + (\eta k / k_m)(\bar{V}/S_A)} \tag{7-48}$$

where the effectiveness factor η is that determined previously (intraphase). The quantity $(k/k_m)(\bar{V}/S_A)$ in equation (7-48) is called the *Damköhler number* (N_{Da}) and expresses the relative rates of the intraphase reaction to external mass transport. A little clearer physical interpretation is obtained, however, if we write equation (7-48) in terms of the ratio of the inter- to intraphase transport coefficients, a quantity termed the *mass Biot number*, N_{Bi_m}. We define

$$N_{Bi_m} = \left(\frac{k_m}{D_{eff}}\right)\left(\frac{\bar{V}}{S_A}\right) \tag{7-49}$$

Then

$$\eta N_{Da} = \eta \frac{k(\bar{V}/S_A)^2}{(D_{eff})(N_{Bi_m})}$$

which is

$$\eta N_{Da} = \phi \frac{\tanh \phi}{N_{Bi_m}} \qquad \text{(slab)}$$

$$\eta N_{Da} = \frac{\phi_s \coth \phi_s - 1}{3 N_{Bi_m}} \qquad \text{(sphere)} \tag{7-50}$$

This gives, for the global rate of reaction in equation (7-48),

$$(-r) = \frac{\eta k C_0}{1 + [(\phi \tanh \phi)/N_{\text{Bi}_m}]} \qquad \text{(slab)} \qquad (7\text{-}51)$$

An extensive discussion of various limiting forms of equation (7-51) has been given by Carberry [J. J. Carberry, *Catalysis Reviews*, 3, 61 (1969)].

Alternative to the derivation of equation (7-51), the isothermal mass balance may be solved analytically with the boundary conditions of equation (7-45). For spherical geometry,

$$\eta_0 = \left(\frac{3 N_{\text{Bi}_m}}{\phi_S^2}\right) \frac{\phi_S \cosh \phi_S - \sinh \phi_S}{\phi_S \cosh \phi_S + (N_{\text{Bi}_m} - 1) \sinh \phi_S} \qquad (7\text{-}52)$$

The effectiveness factor of equation (7-52) is an overall value and is *not* equivalent to the intraphase effectiveness factor used in the derivation of equation (7-51). The Biot number above is defined as $(k_m R_p/D_{\text{eff}})$, where R_p is the particle radius.

Interphase/intraphase selectivity problems under isothermal conditions have also been treated by Carberry. For the important Type III system, the differential selectivity corresponding to equation (7-39) is

$$\Delta_{\text{diff}} = \frac{m_1 \phi_2 \tanh \phi_2}{m_2 \phi_1 \tanh \phi_1} \left(R + \frac{S_i}{S_i - 1}\right) + \frac{S_i}{S_i - 1} \qquad (7\text{-}53)$$

where the effective diffusivities of A and B are the same, and reaction transport parameters are defined on the basis of $(\bar{V}/S_A)$ as the characteristic dimension. The quantities m_1 and m_2 are defined as

$$m_i = 1 + \frac{\phi_i \tanh \phi_i}{N_{\text{Bi}_m}} = 1 + \frac{\phi_i^2 \eta}{N_{\text{Bi}_m}} \qquad (7\text{-}54)$$

Integration of equation (7-53) over a range of conversions gives the analog of equation (7-43):

$$f_B = \frac{S_i}{S_i - 1}[(1 - f_A)^\Psi - (1 - f_A)] + R_i(1 - f_A)^\Psi \qquad (7\text{-}55)$$

where R_i is the initial concentration ratio, (C_{A_0}/C_{B_0}), and Ψ is

$$\Psi = \frac{m_1 \phi_2 \tanh \phi_2}{m_2 \phi_1 \tanh \phi_1} = \frac{m_1 \phi_2^2 \eta_2}{m_2 \phi_1^2 \eta_1}$$

The results of some example calculations with equations (7-53) and (7-55) are given in Figure 7.11.

The other two major types of selectivity are more easily analyzed. For Type II it may be shown, as previously, that if reaction orders are the same, diffusion has no influence on selectivity (or, more precisely, influences both reactions equivalently). For Type I reactions, the differential selectivity is directly determined from the ratio of rates for the two reactions as computed from equation (7-51).

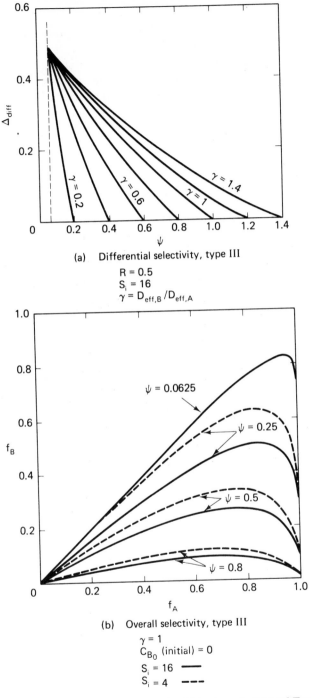

Figure 7.11 Differential and overall selectivities in an isothermal Type III reaction with interphase and intraphase gradients (after J. J. Carberry, *Chem. Eng. Sci.*, **17**, 675 (1962); with permission of Pergamon Press, Ltd., London)

The analysis of activity and selectivity in nonisothermal reactions requires the numerical solution to the mass- and energy-balance equations with boundary conditions of equation (7-45). In nondimensional form for first-order kinetics, these are

$$\nabla^2 f - \phi^2 f e^{-\gamma/\tau} = 0$$
$$\nabla^2 \tau + \beta \phi^2 f e^{-\gamma/\tau} = 0 \tag{7-56}$$

$$\frac{\partial f}{\partial \zeta} = \frac{\partial \tau}{\partial \zeta} = 0 \qquad \zeta = 0$$

$$\left.\begin{array}{c} \dfrac{\partial f}{\partial \zeta} = N_{\mathrm{Bi}_m}(1 - f_S) \\[2mm] \dfrac{\partial \tau}{\partial \zeta} = N_{\mathrm{Bi}_h}(1 - \tau_S) \end{array}\right\} \qquad \zeta = 1$$

where the normalization is with respect to bulk conditions, and N_{Bi_h} is a Biot number for heat transfer, (hR_p/k_{eff}). The effectiveness factor may be written

$$\eta = \frac{3N_{\mathrm{Bi}_m}}{\phi^2}(1 - f_S) \tag{7-57}$$

and from the Prater relationship we have

$$\tau = \tau_S + \beta(f_S - f) \tag{7-58}$$

so

$$\frac{\partial \tau}{\partial \zeta} = -\beta \frac{\partial f}{\partial \zeta}$$

which can be employed to show that

$$\tau = 1 + \beta \mu(1 - f_S) + \beta(f_S - f) \tag{7-59}$$

where

$$\mu = \frac{N_{\mathrm{Bi}_m}}{N_{\mathrm{Bi}_h}}$$

Equation (7-59), expressing τ as a function of f, allows one to reduce the pair of equations (7-56) to a single equation for numerical solution, as described by Weisz and Hicks for the intraphase problem. Some results of effectiveness-factor calculations for first-order reactions are given in Figure 7.12; these indicate clearly the significant effects that boundary-layer resistances have on catalytic activity and behavior. Both plots indicate that the locus and magnitude of the region of multiplicity are altered with changing N_{Bi_m} and N_{Bi_h}. In the top of the figure is illustrated variation with N_{Bi_h} at a fixed ratio, $\mu = 50$, while at the bottom is illustrated variation with N_{Bi_m} with a fixed ratio of 2. Particularly striking are the indicated minima in effectiveness factor occurring at intermediate values of the Thiele modulus. This is associated with the magnitude of the external thermal gradient and corresponds

TABLE 7.3

TORTUOSITY FACTORS FOR DIFFUSION IN CATALYSTS

Catalyst	Technique	ϵ	τ	Reference[a]
Ag/8.5% Ca alloy pelletized from powder	Gas diffusion, 1 atm	0.3 0.04	6.0 ∞	1
Ag pelletized from above powder after removal of Ca by leaching	Gas diffusion, 1 atm	0.6 0.3 0.1	7.5 10 ∞	1
Ni pelletized from commercial powder	Gas diffusion, 1 atm	0.26	6	1
Ag pelletized from powder	Gas diffusion, 1 atm	0.7 0.3	1.7 3.3	2
Sprayed Ag-alloy catalyst, after activation	Gas diffusion, 1 atm	0.68 0.41	2–3.5 16	3
Pelletized boehmite alumina	Gas diffusion, 1 atm	0.34 0.31	2.7 1.8	4
Pelletized Cr_2O_3/Al_2O_3 catalyst	Gas diffusion, 1 atm	0.22	2.5	5
1% Pd on alumina spheres, commercial support	From reaction (liquid phase)	0.5	7.5	6
0.5% Pd on alumina, commercial-type catalyst pellets	From reaction (liquid phase)	0.59	3.9	7
Pelletized from 1 to 8-μ iron powder	Gas diffusion, 1 atm	0.22–0.32	2.6–2.9	8
Harshaw commerical MeOH synthesis catalyst, prereduced	Gas diffusion, 65 atm	0.49	6.9	9
Haldor-Topsøe commercial MeOH synthesis catalyst, prereduced	Gas diffusion, 65 atm	0.43	3.3	9
BASF commercial MeOH synthesis catalyst, prereduced	Gas diffusion, 65 atm	0.50	7.5	9
Girdler G-52 commercial catalyst, 33% Ni on refractory oxide support	Gas diffusion, 65 atm	0.44	4.5	9
Girdler G-58 commercial catalyst, Pd on alumina	Gas diffusion, 65 atm	0.39	2.8	9

[a] 1, C. M. Amberg and E. Echigoya, *Can. J. Chem. Engr.*, **39**, 215 (1961); 2, S. Masamune and J. M. Smith, *Amer. Inst. Chem. Eng. J.*, **8** 217 (1962); 3, G. L. Osberg, A. Tweddle, and W. C. Brennan, *Can. J. Chem. Engr.*, **41**, 260 (1963); 4, N. Wakao and J. M. Smith, *Chem. Eng. Sci.*, **17**, 825 (1962); 5, C. N. Satterfield and S. K. Saraf, *Ind. Eng. Chem. Fundls.*, **4**, 451 (1965); 6, C. N. Satterfield, A. A. Pelossof, and T. K. Sherwood, *Amer. Inst. Chem. Eng. J.*, **15**, 226 (1969); 7, C. N. Satterfield, Y. H. Ma, and T. K. Sherwood, cited in C. N. Satterfield, *Mass Transfer in Heterogeneous Catalysis*, MIT Press, Cambridge, Mass., 1970; 8, J. Hoogschagen, *Ind. Eng. Chem.*, **47**, 906 (1955); 9, C. N. Satterfield and P. J. Cadle, *Ind. Eng. Chem. Fundls.*, **7**, 202 (1968). *Source:* After C. N. Satterfield, *Mass Transfer in Heterogeneous Catalysis*, © 1970; with permission of MIT Press, Cambridge, Mass.

density of ρ_b (wt/volume) we may write

$$\bar{R} = \frac{2V_g}{S_g} = \frac{2\epsilon}{S_g \rho_b} \tag{7-68}$$

The following procedures may then be used to obtain D_i pertinent to the type of diffusion mechanism that predominates:

1. *Bulk diffusion*—independent of $\bar{R}$. Values obtained from tabulated data, interaction theory (J. O. Hirschfelder, C. F. Curtiss, and R. B. Bird, *Molecular Theory of Gases and Liquids*, Wiley, New York, 1954), or semitheoretical correlations [for example, J. C. Slattery and R. B. Bird, *Amer. Inst. Chem. Eng. J.*, **4**, 137 (1958)].

2. *Knudsen diffusion.* For diffusion of a molecule A, an analog to the result for regular cylindrical capillaries is

$$D_{K_A} = \left(\frac{2\bar{R}}{3}\right)\left(\frac{8kT}{m_A}\right)^{1/2} \qquad \text{(molecular units)}$$

or

$$D_{K_A} = 9700\bar{R}\left(\frac{T}{M_A}\right)^{1/2} \ \text{cm}^2/\text{sec} \tag{7-69}$$

where $T = {}^\circ\text{K}$, M_A is the molecular weight, and $\bar{R} = \text{cm}$.

3. *Transition diffusion.* This is a bit more complicated than the two limits, since the theoretical result for D_i does not fit exactly into the form of equation (7-65). For binary diffusion of A in B, the following has been shown to apply [R. B. Evans, G. M. Watson, and E. A. Mason, *J. Chem. Phys.*, **33**, 2076 (1961)].

$$D_i(\text{transition}) = \left(\frac{1 - \alpha x_A}{C\mathfrak{D}_{AB}} + \frac{1}{CD_{K_A}}\right)^{-1} \tag{7-70}$$

where C is total concentration, x_A the mole fraction of A, $\mathfrak{D}_{AB}$ and D_{K_A} bulk and Knudsen diffusion coefficients, respectively, and α a quantity related to the ratio of fluxes of B to A, N_B/N_A, as

$$\alpha = 1 + \frac{N_B}{N_A} \tag{7-71}$$

The diffusivity is thus dependent on concentration and flux ratio, and strictly speaking we cannot use a theory of catalyst effectiveness based on constant diffusivity. This, however, is probably building more detail into the theory than it is worth; an adequate approximation may be obtained from the limit of equation (7-70) for constant-pressure conditions and $N_B = -N_A$. Then

$$D_i(\text{transition}) = \left(\frac{1}{\mathfrak{D}_{AB}} + \frac{1}{D_{K_A}}\right)^{-1}$$

For small values of $\sqrt{M}$, that is, when k is small, the process approaches the limit of pure physical diffusion, $\lambda = 1$, for all values of rq. At high values of $\sqrt{M}$, an asymptotic limit is approached, the value dependent on rq, which is the solution given by equation (7-82). Finally, for $rq = \infty$, the solution becomes that for a pseudo-first-order reaction, essentially the result of Hatta. Note that determination of the value of $\sqrt{M}$ requires a value for the mass-transfer coefficient in the absence of chemical reaction.

b. Diffusion and reaction: penetration theory

A popular alternative to the film theory for mass transfer has been the penetration theory, in which the mass transfer is viewed as unsteady transport into an element of fluid which is periodically removed from the interface and replaced by a fresh element. If no chemical reaction occurs, the average mass transfer rate over a given contact time θ is

$$N_A = 2\sqrt{\frac{D_A}{\pi\theta}}(C_{A_t} - C_{A_L}) \tag{7-84}$$

so the mass-transfer coefficient according to penetration theory is

$$k_m = 2\sqrt{\frac{D_A}{\pi\theta}}$$

In effect, this approach swaps the arbitrary thickness, L, of the film theory for the arbitrary contact time, θ. Enhancement factors for a number of cases have been determined in terms of this theory. For a first-order reaction [P. V. Danckwerts, *Trans. Faraday Soc.*, **46**, 300 (1950)]

$$\lambda = \frac{1}{2}\left[\sqrt{\frac{\pi}{k\theta}}\left(\frac{1}{2} + k\theta\right)\right]\mathrm{erf}\sqrt{k\theta} + \exp(-k\theta) \tag{7-85}$$

where k is the first-order rate constant and erf is defined in the usual way.

For second-order reactions, the asymptotic solution for infinitely rapid reaction [P. V. Danckwerts, *Trans. Faraday Soc.*, **46**, 701 (1950)] is given by the awkward parametric form

$$\lambda = \frac{1}{\mathrm{erf}(\sigma)} \tag{7-86}$$

where σ is obtained from the solution of

$$q\sqrt{r} = \frac{1 - \mathrm{erf}(\sigma/\sqrt{r})}{\mathrm{erf}(\sigma)\exp[\sigma^2(1 - 1/r)]} \tag{7-87}$$

Finally, when finite rates of reaction are involved in the second-order case, Gilliland et al. [E. R. Gilliland, R. F. Baddour, and P. L. T. Brian, *Amer. Inst. Chem. Eng. J.*, **4**, 223 (1958)] proposed that the following correlation, modeled on equation (7-83), could be used as an approximation to results obtained numerically:

$$\lambda' = \frac{\sqrt{M}\,\sqrt{1 - (\lambda' - 1)/(\lambda - 1)}}{\tanh\left[\sqrt{M}\,\sqrt{1 - (\lambda' - 1)/(\lambda - 1)}\right]} \tag{7-88}$$

where λ is determined from equations (7-86) and (7-87) and $\sqrt{M}$ is as defined for equation (7-83). Brian et al. [P. L. T. Brian, J. F. Hurley, and E. H. Hasseltine, *Amer. Inst. Chem. Eng. J.*, **7**, 226 (1961)] found that the approximation provided by equation (7-88) becomes only fair for certain ranges of $\sqrt{M}$ and λ, and have published correction factors for the approximation in terms of these parameters.

In general, penetration theory results differ significantly from film theory results only when the diffusivity ratio r differs significantly from unity, so the additional complexity involved in the solutions of equations (7-86) to (7-88) should reasonably restrict their use to this situation. For extensive discussions of the theory of diffusion and reaction in gas/liquid systems, both in terms of film and penetration models, the reader is referred to the works of Astarita (G. Astarita, *Mass Transfer with Chemical Reaction*, Elsevier, Amsterdam, 1967) and Danckwerts (P. V. Danckwerts, *Gas–Liquid Reactions*, McGraw-Hill, New York, 1970).

c. Selectivity factors in gas/liquid reactions

The alteration of selectivity due to transport in gas/liquid reactions is less commonly encountered than for gas/solid reactions. Basically, the reason for this is that the reaction must essentially occur completely in the film for transport to alter selectivity. If the reaction occurs predominantly in the bulk phase, concentrations of reactants and products are observable and selectivity is simply determined by the reaction kinetics in the homogeneous phase.

Selectivity in gas/liquid systems has been examined in detail by van de Vusse [J. G. van de Vusse, *Chem. Eng. Sci.*, **21**, 631, 645 (1966)] for the consecutive reactions

$$A + B \longrightarrow C \qquad (k_1)$$
$$A + C \longrightarrow D \qquad (k_2)$$

which reduces to a Type III scheme when A (from the gas phase) is present in considerable excess. Schemes such as the above are typical of successive substitution reactions, for example the chlorination of paraffins (*n*-decane was studied by van de Vusse) or aromatics such as benzene. The individual reaction steps are second-order and so may be analyzed in terms of the general theory presented in the preceding section.

In terms of the reaction parameter $\sqrt{M}$, two limiting regions of behavior may be identified. For $\sqrt{M} < 0.5$, the reaction takes place in the bulk liquid phase and, while the rate of reaction may be affected by the mass transport, the selectivity is unaffected. For $\sqrt{M} > 2$, the concentration of A is essentially depleted within the film and no reaction occurs in the bulk liquid phase. Whether there is a significant gradient of B within the film, and hence whether there is an effect on selectivity, depends entirely on the relative concentrations of A and B as given by the ratio (C_{B_L}/C_{A_i}). For large values of the ratio, $(C_{B_L}/C_{A_i}) \gg \sqrt{M}$, the concentrations of B and C in the film are nearly constant; the rate is affected but the selectivity is not changed. In this case the rate is given by

$$(-r_A) = N_A a = aC_{A_i}\sqrt{k_1 D C_{B_L}} \tag{7-89}$$

where D is a diffusion coefficient and a is the gas/liquid specific interfacial area. For values of the ratio $(C_{B_L}/C_{A_i}) \approx \sqrt{M}$ and smaller, reactant B is significantly depleted in the film and both rate and selectivity are affected. A summary of this analysis is given in Table 7.6.

TABLE 7.6
MASS-TRANSPORT EFFECTS ON RATE AND SELECTIVITY

	$\sqrt{M} > 2$	$\sqrt{M} < 0.5$
$\dfrac{C_{B_L}}{C_{A_i}} \gg \sqrt{M}$	Effect on rate, none on selectivity	No effects on selectivity
$\dfrac{C_{B_L}}{C_{A_i}} \approx \sqrt{M}$	Effect on rate and selectivity (A and B profiles in film)	If $k_1 C_{B_L} \ll \dfrac{D}{L}(a)$, chemical reaction rate $= (-r_A) = k_1 C_{A_i} C_{B_L}$
$\dfrac{C_{B_L}}{C_{A_i}} \ll \sqrt{M}$	Effect on rate and selectivity (location of reaction plane)	If $k_1 C_{B_L} \gg \dfrac{D}{L}(a)$, physical transfer rate $= (-r_A) = a\left(\dfrac{D}{L}\right) C_{A_i}$

Now in the cases where selectivity is affected, this will be expressed in the ratio (C_{C_L}/C_{B_L}), which will vary with extent of reaction. Van de Vusse found that a satisfactory correlation between the variation of C_{C_L} and C_{B_L} could be written

$$\frac{dC_{C_L}}{dC_{B_L}} = \alpha\frac{C_{C_L}}{C_{B_L}} - \beta \tag{7-90}$$

With the initial condition that $C_{C_L} = 0$ for $C_{B_L} = C_{B_L}^\circ$, this may be integrated to

$$\left(\frac{1-\alpha}{\beta}\right)C_{C_L} = C_{B_L}^\alpha - C_{B_L} \tag{7-91}$$

Now, either in the case that the kinetics of the sequence do not depend on the concentration of A or that $C_{A_i} \gg C_{B_L}$, the mass conservation equations may by integrated analytically to obtain the following values for α and β:

$$\alpha = \left(\frac{k_2}{k_1}\right)^{1/2} \frac{\tanh \sqrt{M_2}}{\tanh \sqrt{M_1}} \tag{7-92}$$

where

$$\sqrt{M_1} = \frac{\sqrt{k_1 D C_{A_i}}}{k_m}$$

$$\sqrt{M_2} = \frac{\sqrt{k_2 D C_{A_i}}}{k_m}$$

and

$$\beta = \frac{k_1}{k_1 - k_2}(1 - \alpha) \tag{7-93}$$

Without diffusion limitation $\sqrt{M_1} < 0.3$ and α and β are given by

$$\alpha = \frac{k_2}{k_1} \qquad \beta = 1 \tag{7-94}$$

For strong diffusion limitation $\sqrt{M_1} > 2$ and

$$\alpha = \sqrt{\frac{k_2}{k_1}} \qquad \beta = \frac{1}{1 + \sqrt{k_2/k_1}} \tag{7-95}$$

For values of $\sqrt{M_1}$ intermediate between these limits, the mass conservation equations must be solved numerically, as is also required if $C_{A_i} < C_{B_L}$. The results of a typical calculation for this system are given in Figure 7.19 for a reaction system of strong diffusion control, $\sqrt{M_1} = 6.3$. Here the maximum yield $(C_{Cmax}/C_{B_L}^{\circ})$ is plotted as a function of the concentration ratio $(C_{B_L}^{\circ}/C_{A_i})$. The limiting value of yield for $(C_{B_L}^{\circ}/C_{A_i}) = 0.1$ is computed from equations (7-91) and (7-95) for flat A profile and strong diffusion, while the limit of no diffusion effect is computed from equation (7-91) with α and β given by equation (7-94). It can be seen that the no diffusion limit is approached as $(C_{B_L}^{\circ}/C_{A_i})$ becomes much larger than $\sqrt{M_1}$, as was indicated in the previous qualitative discussion and shown in Table 7.6.

The dependence of selectivity and yield on concentrations is typical of reactions in gas/liquid systems, and makes generalization of results such as these quite difficult. The particular case treated here, rapid bimolecular reaction, is, however, the most important instance in which selectivity alteration may occur in gas/liquid reactions.

Studies of selectivity in gas/liquid reactions directly concerned with Type III reactions have been reported by Bridgewater [J. Bridgewater, *Chem. Eng. Sci.*, **22**, 185 (1967)] using the film theory for mass transfer, and by Szekeley and Bridgewater [J. Szekeley and J. Bridgewater, *Chem. Eng. Sci.*, **22**, 711 (1967)] using a penetration theory analysis. Interestingly, it is shown in these two papers that some difference obtains in the results for selectivity provided by these two approaches for the Type III reaction. In the simpler case of a single direct reaction, where one is concerned only with conversion (or rate

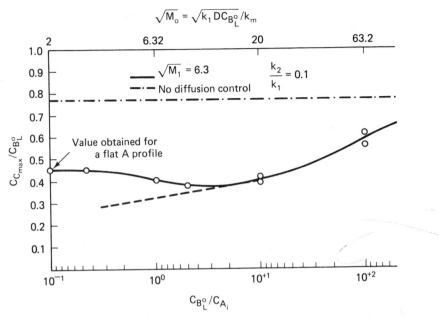

Figure 7.19 Influence of diffusion on the yield in a consecutive second-order gas/liquid reaction (after J. G. van de Vusse, *Chem. Eng. Sci.*, **21**, 631 (1966); with permission of Pergamon Press, Ltd., London)

of mass transfer), there is little difference in the film or penetration theory results.

7.3 First-Level Applications to Reactor Design

When one is concerned with the design or modeling of a reactor in which more than one phase is involved, a very fundamental question presents itself at the onset. Shall we retain the models we have employed to date and use them in a pseudo-homogeneous approximation, or shall we expand them into separate balances for each of the two phases? Certainly in terms of the esthetics of physical and chemical modeling, the latter is probably always the preferable course. On the other hand, there is no point in putting more detail into a model than is necessary to attain the desired design objectives. Since there is probably an order of magnitude more effort involved in the implementation of two-phase reactor models, the pseudo-homogeneous representation remains a useful and important means for reactor modeling. It is important to know, of course, when this approach will not be adequate for the needs at hand, as we have discussed in chapters 5 and 6. The incorporation of transport effects in reactor models using the pseudo-homogeneous approach can be accomplished on the basis of the theories detailed here. We can view this as sort of a first-level approximation to reality which may be entirely sufficient for some types of design problems and useful in the construction of exploratory models for other situations.

Application of the theory in this case is quite simple, in fact, since we have been careful to define quantities such as the effectiveness factor or the enhancement factor in terms of the rate under observable conditions. Consider as an example a PFR catalytic reactor model in which we wish to include possible diffusion limitations on the catalyst activity. We start with the design equation written in the following form:

$$-\frac{u}{\epsilon}\frac{dC}{dz} = (-r)_0 \qquad (7\text{-}96)$$

in which u is the linear velocity based on the empty tube cross section, ϵ the bed porosity, and $(-r)_0$ the observed rate of reaction. Now we have available a model for the intrinsic kinetics of the reaction, say $(-r) = kC$, and the observed rate may be determined from this using an effectiveness factor appropriate for the situation. For example, if the reactor is isothermal and we are concerned with intra- and interphase mass-transport limitations,

$$\eta = \left(\frac{3N_{\mathrm{Bi}_m}}{\phi_s^2}\right)\frac{\phi_s \cosh \phi_s - \sinh \phi_s}{\phi_s \cosh \phi_s + (N_{\mathrm{Bi}_m} - 1)\sinh \phi_s} \qquad (7\text{-}52)$$

for spherical catalyst geometry and the reactor model is

$$-\frac{u}{\epsilon}\frac{dC}{dz} = \eta kC \qquad (7\text{-}97)$$

This is a simple, almost trivial, example but it would not have been so if we had not been so meticulous in our treatment of diffusion and reaction in gas/solid systems. In effect, the model of equation (7-97) is a representation of the following two-phase model:

$$-\frac{u}{\epsilon}\frac{dC}{dz} = k_m a(C - C_0') \qquad (7\text{-}98)$$

and

$$D_{\mathrm{eff}}\left[\frac{1}{r^2}\frac{d}{dr}\left(r^2\frac{dC'}{dr}\right)\right] - kC' = 0 \qquad (7\text{-}99)$$

with

$$
\begin{aligned}
C &= C_0 & z &= 0 & &\text{[equation (7-98)]} \\
C' &= C_0' & r &= R_p \\
\frac{dC'}{dr} &= 0 & r &= 0 &
\end{aligned}
$$
  [equation (7-99)]

where $k_m a$ is a mass-transfer coefficient per unit volume, C the bulk gas-phase concentration in the reactor, C' the concentration within the catalyst phase, and r the radial dimension of the catalyst. It is left as an exercise to the reader to determine which of these alternative model representations would be easier to use as an approximation to the two-phase design problem.

Section 7.2

a	gas-liquid interfacial area, equation (7-89)
a_0	$(k/D)^{1/2}$
C	concentration in liquid phase, moles/volume
C_i, C_L	concentrations at interface or in bulk, moles/volume
C_{A_i}, C_A, C_{A_L}	concentrations of A at various points, moles/volume
C_{B_i}, C_B, C_{B_L}	concentrations of B at various points, moles/volume
$C_{B_L}^{\circ}$	initial concentration of B at L, moles/volume
$C_{C_{max}}$	maximum concentration of C, moles/volume
D, D_i	diffusion coefficients in the liquid, phase, area/time
k_m^*	$k_m \lambda$
L, L', L''	film thickness, length
M	parameter, equations (7-83) and (7-88)
$M_{1,2}$	$(k_{1,2}DC_{A_i})^{1/2}/k_m$
r, q	parameters, equation (7-83)
α, β	parameters, equation (7-90)
θ	contact time
λ, λ'	enhancement factor, equations (7-82), (7-85), (7-88)
ν	moles B reacted per mole A, equation (7-83)
σ	parameter, equation (7-86)

Section 7.3

a	mass transfer, area/volume
C, C'	fluid phase and intraparticle concentrations, moles/volume
C_0'	concentration at particle surface, moles/volume
$(-r)_0$	overall (observed) rate of reaction, moles/volume-time
u	linear velocity, length/time
ϵ	bed porosity

So I leave it with all of you:
Which came out of the opened
door—the lady or the tiger?

Frank Richard Stockton

Author Index

Index

Subject Index